SCIENCE AND ENGINEERING POLICY SERIES

General Editors Sir Harrie Massey
Sir Frederick Dainton

The humane technologist

DUNCAN DAVIES
TOM BANFIELD
RAY SHEAHAN

OXFORD UNIVERSITY PRESS 1976

Oxford University Press *Ely House, London, W.I.*

Glasgow	Delhi
New York	Bombay
Toronto	Calcutta
Melbourne	Madras
Wellington	Karachi
Cape Town	Kuala Lumpur
Ibadan	Singapore
Nairobi	Jakarta
Dar es Salaam	Hong Kong
Lusaka	Tokyo
Addis Ababa	

ISBN 0 19 858325 7

Printed in Great Britain
by Billings & Sons Ltd., Guildford and Worcester

Preface

This book is first of all intended to help technologists approach their increased and increasing range of problems, now extending well outside the subjects in which they were trained and educated, and into economics, politics, philosophy, and most aspects of human behaviour. Attacks on these problems (and opportunities) call for collaboration, on terms of close mutual understanding, with non-technologists, for whom therefore we have also tried to write, to enable them to understand technological difficulties, share perceptions, and use common or compatible methods. In the past, there have been serious linguistic difficulties: for example, technologists have preferred to use numbers and mathematical arguments, and non-technologists words and conceptual arguments. Mutual understanding is improving, but there is a long way to go. To present such a broad range of topics and proposals to such a broad range of readers involves the risk of boring the technologists by explanation of matters obvious to them, but without which their non-technological co-readers would be confused or lost. We therefore plead for patience on both sides, because it seems necessary and valuable to be comprehended by both, even if the compromise in content leads to a treatment that is not entirely satisfactory to either.

Apart from content, there is also a problem of style. Technologists are accustomed to writing that is extensively, authoritatively, and explicitly illustrated, whose conclusions are highly probable and commanding, though often correspondingly limited. We cannot offer the satisfactory certainty and authority of a book on mechanical or electrical engineering, or on computer systems and logic, and our field is so large that really adequate illustration would run to several volumes. It may be that a series of such volumes should be written later, but as a start we have sought to be readily readable, sacrificing rigour in consequence. We hope that a reader can complete a first perusal, skipping difficulties, during a train journey from London to Glasgow or

Edinburgh, or a westward flight across the Atlantic, America, or Australia, allowing an hour for a meal. Each of these journeys gives between three and five hours without interruption.

In this book we are seeking to build on foundations laid in an earlier book on technology and economics, and on the studies in technological economics and science policy that were started in various universities in the middle 1960s, when that book was written. Both books arise from practice and pragmatism, not theory or principle, for we are ourselves in daily need of means for progressively improving the linkages between technology and its users and neighbours. We then wrote on the basis of the needs of a substantial group of young industrial laboratory scientists who were seeking to make and develop inventions of the right kind. Now we are writing as part of the job of knitting together the investment pattern of the same international company (making products ranging from pharmaceuticals to heavy chemicals, plastics, and textiles) and of partitioning available cash and other resources between extending factories and plants, acquiring interests in related businesses, and creating new opportunity through research. All these activities have to be paid for with the same kind of money from the same cash trading margin, and all have to be looked after by the same types of able and energetic people and linked with customers and communities in different countries.

Then, we were able to formulate a new approach to a single academic and practical discipline—that of economics—that was intended to be, and has turned out to be, particularly acceptable to technologists. Microeconomics—the assessment of cost and benefit at the level of the project and the portfolio of projects—can be immediately used by researchers, developers, factory managers, and technical marketers to make better decisions about priorities and timing. It was not until a decade ago a highly regarded or usual approach to economics. But it involves money as a scale of measurement of resources as diverse as raw materials, labour, energy, and investment; and as such it is at once comprehensible to the technologist. In this book, the task is harder and the recommendations are less definite; it is only about the utility of a method—that of modelling and simulation—that we are confident. Consequently, this time we are not suggesting *exactly* what to go out and do next Monday; we are only saying what kind of method will certainly help.

We have dissected the problem so as to help with its study. Chapters 2 and 3 describe the problems, nature, and growth-pattern of technology. Chapter 4 describes modelling methods in more detail. Chapters 5 to 9 apply the planner's procedure (describe the present position, say where present trends will lead, and then discuss policy options and means for going elsewhere) to means for dealing with each important constraints in turn: finance, resource limitation, human willingness and unwillingness, and the law and the environment.

Technology is now pursued by some, and opposed by others, with fervour such as has more usually been applied to religious doctrine or dogma. We recognize that many changes afterwards accepted as good have required such fervour, and that others, retrospectively condemned as bad, needed equal fervour to stop them because of their momentum. In this book we have tried not be fervid either way, not therefore guaranteeing never again to take part in a revival meeting, nor condemning fervour. The satisfactory conclusion of a technological contract may need people who are moved by fire and endowed with the gift of tongues, but it also requires a majority of pragmatic participants (for whom we are primarily writing) exercising cool judgment.

What has emerged is the result of a great deal of discussion and of practical trial. Among our own colleagues we would like to offer grateful acknowledgement to Dr. A. A. L. Challis, Dr. P. V. Youle, Mr. D. S. Binsted, Mr. A. R. Burgess, and many others, including Mrs. C. Savage, Mrs. I. Cooke, Mr. R. Ellis, and Mr. R. Pryke, who helped with typing, bibliography, and indexing. In other companies we have learnt much from Mr. M. C. Throdahl of Monsanto, Mr. P. W. Longland of B.A.T., and Dr. J. B. P. Williamson of G.K.N. In the academic world, our interlocutors have been Professor F. R. Bradbury and the late Dr. T. L. Cottrell of Stirling, who together turned technological economics into a coherent subject, and Professor C. H. Vereker. Among government servants, we have been especially helped by our contacts with Dr. J. R. Price and Dr. J. A. Allen of C.S.I.R.O., Australia. Finally, we had a great deal of helpful support, comment, and question from Sir Frederick Dainton, Chairman of the U.G.C. and a General Editor of this series; in particular, he asked the important question 'what is a technologist?' Although questions of lexicography or semantics are often dull or frustrating, this one is not, as we set out to explain in the first chapter.

London, Cleveland, and Widnes   D.S.D.
January 1976   T.L.B.
                                                                   R.J.S.

Acknowledgements

Our thanks are due to the following authors and publishers for permission to use material as follows: the Boston Consulting Group Inc. for Fig. 4.2 and Table 3.1; Dr. F. R. Bradbury and the Edinburgh University Press for Fig. 6.1, reproduced from *Words and numbers* (1969); the Central Statistical Office for Fig. 6.4 and Tables 5.6–10; Professor P. Cloud and Sinauer Associates for Table 5.3 from *Environment, resources, pollution and society*, edited by William W. Murdock; Edmund Faltermayer and Time-Life Inc. for Table 5.12, from *Fortune;* Her Majesty's Stationery Office for Fig. 6.4, from *Financial Statistics 1975*, and Tables 3.1, 5.6–10 from *Annual Abstract of Statistics 1972*; Imperial Chemical Industries Ltd. for Fig. 6.3 and Table 5.13; Penguin Books Ltd. for Table 7.1; Dr. F. Roberts and the U.K. Atomic Energy Authority for Table 5.1; the Royal Society for Figs. 3.2, 3.3; Table 5.5 from *Proc. R. Soc.*; the Universities Central Council for Admissions for Table 3.2.

Contents

1

Introduction: What is a technologist?

The study of people does not feature prominently in technological education. Nor do people always behave as economic animals; and even when they do, the economic perceptions and aspirations of interdependent groups may be incompatible as initially stated. One man's wage increase creates his neighbour's price increase, which in turn contributes to further wage claims. This is only one of the complex human systems on which technologists depend. Others are to be found within the factory itself: superficially, the job specialization of the assembly line reduces costs by simplifying training requirements and necessary skills. In time, it fails to satisfy the psychological and social needs of those involved, who rebel in various ways, and who soon find that small groups can wield great power because of the wide range of operations that stop if they strike or go slow. Again, factories annoy their neighbours because of noise, smell, smoke, or river pollution and, as affluence increases, citizens become unresponsive to arguments that power stations *must* emit sulphur dioxide if people want cheap electricity or, more generally, that 'where there's muck there's brass'. Muck and atmospheric sulphur dioxide become regarded as basically intolerable social evils which, like slavery, cannot be justified by arguments about economic benefits.

Human systems thus have increasing importance, and a technologist who understands them and works with, rather than against, them will do better. This calls for a new and complex kind of judgement, for he will have competitors facing the same problems but able to decide differently and perhaps more cheaply. They may have access to immigrant labour with a recent memory of poverty or oppression and thankful to do jobs that his own fellow-countrymen may spurn. They may be able to put their effluent out into a tideway, while he is far inland. And their governments may tax more or less heavily than his. With such complex constraints, living and competing successfully is therefore not easy.

Introduction: What is a technologist?

Not many books have been written about the conduct of a complex task: most books are about situations, or individuals, particular skills, or particular assemblies of fact and observation. Anthony Trollope's *The Warden* and Angela Thirkell's *The Headmistress* are not books setting out to help wardens and headmistresses (although, indirectly, they may do just this). Izaak Walton's *The Compleat Angler* comes nearer, although he is describing a narrower skill. Yet there is one work that performs just such a task as we are attempting. This is Machiavelli's *The Prince.* Because this work has been misunderstood (often without being read completely or carefully) and because its pragmatism has been seen as cynical rather than in pursuit of public good, it is important that we seek to avoid these hazards.

The Prince has simple messages, mainly about human systems and about the constraints that they impose. Mercenary armies are useless if they lose, and can fleece you if they win; he who employs them is on a hiding to nothing. Citizen armies, however, can be stirred to action only in real emergency, and the prince who rises to power as the saviour in time of trouble faces his real problems when he has conquered the immediate enemy and now has to justify his retention in office in time of tranquillity. It is at this time that an hereditary prince, enjoying traditional loyalties, is at special advantage. And so forth. Antony Jay, in *Management and Machiavelli,* has pointed out that modern human systems, though often changed in name, are usually unaltered in substance: thus Machiavelli's observations are accurate and relevant today when the vocabulary has been properly translated and when one has identified, in an international company, who are the courtiers and who are the barons, and how the prince stands in relation to both. The Head of the U.K. Civil Service, hardly a ruthless cynic, has recently recommended such perceptions to his colleagues.

The Prince was also written about small communities, of thousands rather than millions. It is for this reason that it is especially encouraging to the technologist, whose immediate world can, if he does well, have much of the collusion and intimacy of an Italian city-state. Equally, it can be torn apart by disputes between Montagues and Capulets. But the technologist's constraints are much more multifarious and complex than those of a prince, and so our book is inevitably more complex than *The Prince.* Further, it has been written by those still in employment, and not yet banished to their estates for insubordination; so it may well be less detached. And although Machiavelli was writing in political terms, there is no reason to suppose that, given modern

2

numeracy and approaches, his pragmatism cannot be translated into economic and social langage.

We cannot advise in detail, because of the diversity of the jobs that technologists do; we can, however, indicate the lines along which typical problems can be attacked. And although this is not a rigorous book we shall, our of sheer force of habit, occasionally use numbers. At the end, the criterion is simple: have we increased your chances of putting a good idea to effective social use? If not, then we have failed. Our success, incidentally, may be indirect. If we make a technologist laugh at a problem about which everyone has been deadly serious, it may well be that we have helped to improve his human system, for he may be able to get others to laugh also. No human system stands a better chance of sinking its differences than a collection of people laughing at the absurdity of the issue about which they have just been fighting. Technologists are often over-serious; so some cartoons and caricatures may possibly do more good than works of the usual dedicated alleged accuracy. It is on this basis that we hope that sociologists, who are often suspicious of practical pragmatism, will discuss our problems with us. We are, after all, living in the same world.

The Prince, though written for a prince, has been read with benefit and enjoyment by many who are not princes, and this is a further reason for regarding it as a suitable model. But an even more important parallelism relates to viewpoint. Niccolo Machiavelli is unjustly regarded, mainly by those who have not read his works, as a ruthless cynic, and this false perception underlies the current usage of the adjective 'machiavellian'. In fact, his approach was pragmatic, and aimed (by the standards of his time) at the good of the state and its inhabitants. When he recommended measures that we might now regard as brutal, it was in pursuit of countervailing and greater general benefit; he was not afraid of risks or trade-offs. And although he may be a little sardonic, he is not particularly cynical.

Most people have a clear general view that a technologist serves social ends by the use of science where this is available, and craft or empiricism where science is lacking. His output may take the form of goods (such as machinery or materials) or systems and services (such as a scheme for increasing the traffic that can safely be handled by an airport, a road network, or a railway system). Usually, he employs known methodology, but where this is insufficient he improvises into the unknown; sometimes he generates new science in so doing. Thus, he has areas of activity in common with the scientists, the traders, and the craftsmen and artists. He has not always had help from science, and drew much initial strength from craft and tradition, on which primitive medicine and metallurgy depended. He has always needed perceptions about market and social needs, often of a more imaginative kind than most traders gave him. (Henry Ford's vision of the massive acceptability of cheap automobiles, and I. K. Brunel's views of railways and great ships, went far beyond orthodoxy.) And although he usually employs logical methods, he is often led by artistic perceptions and produces artistic results in the shape of attractively shaped and sited bridges and buildings, graceful aerodynamic objects, and intellectually pleasing schemes and strategies. Although (as is discussed in Chapter 3) technology has hitherto needed a progressively increasing rate of saving and capital spending, this increase may now slacken and technology's linkage with postponement of consumption may prove to be complex. Similarly (as is discussed throughout the book), specialism, which has so far been technology's central procedure for making problems soluble and opportunities approachable, seems to have its limits, so that technologists are not inherently specialists.

This book recommends models; so perhaps a model will help here. Human activity can be represented as the integration of ideas, material objects, and the lives of people, and differs in the proportions of the mix. The scientist, at one end of his work (as in the study of astronomy, electricity, or matter), is concerned only with ideas and things: his satisfaction can be the intellectual contemplation of a correlation or a model that fits (such as relativity or quantum theory). The trader, at his most orthodox, brings together things and people, with no ideas. The artist, who can occasionally work with ideas alone, is more usually concerned with ideas and people, and some concern with things. Thus, we can draw a triangular Venn diagram (Fig. 1.1), at whose corners we represent

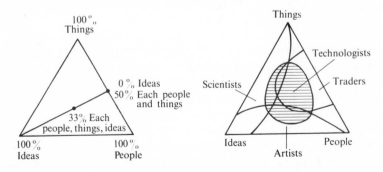

Fig. 1.1. Human activity diagram.

total concern with things, ideas, or people, along whose edges we represent combinations of these elements two at a time, and in whose centre we represent an equal mix of all three. (This form of representation, which depends on the geometric fact of the constancy of the sum of the perpendiculars from any point in an equilateral triangle on to the three sides, is often used in physical chemistry and elsewhere.) The activities of the scientist, artist, and trader might then be shown; the technologist occupies the middle, overlapping with all.

It may be that the technologist's most characteristic skill proves to be his use of models to help study problems and opportunities as systems. In this respect, the *Shorter Oxford Dictionary's* basic definition of 'technology' as 'systematic treatment' now seems more appropriate than its derivative current usage definitions of 'a discourse of treatise on an art or arts; the scientific study of the practical or industrial arts; practical 1859; the terminology of a particular art or subject; technical nomenclature 1658.' *Sancta simplicitas*.

2

The new problems of the technologist

1. The technological contract

Rousseau began his essay on the Social Contract with the following paragraph:

> Man is born free; and everywhere he is in chains. One thinks himself the master of others, and still remains a greater slave than they. How did this change come about? I do not know. What can make it legitimate? That question I think I can answer.

Precisely parallel statements and questions are appropriate today about technology and about the relationship, contractual or otherwise, between a technologist and the community of which he is a part. Rousseau was concerned to advance the novel idea that an individual should, in the eighteenth century, himself become concerned, as a participant, in the formation and formulation of the contractual law to which he would then subject himself. Absolute law and absolute contractual relationship had become a matter of history and should be replaced by a democracy calling for a sensitive appreciation by all individuals of the reasons for social organization and order.[1, 2, 3] Similarly, the technologist, previously concerned primarily with the contracts affecting the creation of new things and making them work, should now, in the late twentieth century, face the need to consider the whole of society and its workings. His work, formerly peripheral, has become sufficiently central to affect the very structure of society and its organization.

It is undoubtedly correct to say that technology was born free, had an unfettered infancy, and is everywhere in chains. At one time the technologist was expected to do no more than make inventions, reduce them to practice, and provide the means for their operation.[4] His wares were much sought after, and he could concentrate all his attention on the engineering and, later, on the underlying science.[5, 6] Then, as his activities spread, they began to interact and to reach a scale that changed whole patterns of existence. Economies of scale led to large factories employing many people; the product of one was the raw material for another, so that interdependence grew. Towns became cities; rivers that could accept the waste from the smaller communities were sometimes poisoned unacceptably by the waste from the larger ones.[7] Capital intensity made manufacture more closely dependent on the banker and on society's willingness to save the necessary resources. Specialization in production operations meant that a small number of people, astride a long chain of manufacture employing many thousands in total, wielded great power and could, at will, use it to considerable effect.

Although the harm—and potential harm—arising from technology has frequently been exaggerated in recent times,[8] so was the benefit sometimes exaggerated in earlier times. Even allowing for the swings of this argument, there is no doubt that the technologist now needs a more highly developed knowledge of markets, people, and society in order to take good decisions and maintain a reasonable and acceptable overall technological contract with the people among whom he lives and whose needs he helps to serve. It is no longer sufficient for him to present his work, ask 'Is it yet good enough and cheap enough?' and, if not, return to the workshop confident that a little more effort will do the trick. He must have thought out and answered questions about cash demands for investment,[9] the acceptability of the jobs that will have to be done, about possible toxicity or pollution, and so forth. These chains have only quite newly become so irksome and heavy. The issues have, of course, been there for several decades, but could previously be dealt with tactically, and often after the factory had been built and was operating. But, just as man himself must accept the chains implicit in Rousseau's social contract if his society is to be a good and stable one, so must the technologist. To live with constraint and to avoid kicking over the traces is a great art, but once learnt it makes progress far more rapid. It is such understanding, and the consequent generation of rapid and acceptable development, that is the subject under discussion now. Whether or not it will use effective or fair methods, the United Sates has an Office of Technology Assessment,[10] and bodies such as this can be expected to be assiduous in formulating the public demand that is part of the technological contract. Because the whole subject is so new, it is hardly surprising that it is a source of bewilderment, resentment, and conflict.[11, 12] Technologists, accustomed to esteem, are finding themselves criticized on grounds that they have not studied, and society, accustomed to technologically based opportunity, is finding itself grappling with these technologically generated problems.

7

Technological education has little to say about people and their responses and preferences: the available time is fully taken up with the properties of matter, the laws of motion, the behaviour of radiation, and other aspects of the inanimate world. And although some technologists come to accept that the proper study of mankind is man, linking this perception with their work, many do not. Indeed, it is not necessary that all should be skilled in human affairs, but those who are not must accept constraint and regulation from those who are, some of whom in turn must have a deep and proper understanding of technology.[13, 14]

There is thus insufficient understanding, as yet, to write definitively on the best lines for designing and operating technological contracts, linking new inventions and procedures with the communities they are meant to benefit. It is not very useful to define the problems and academically leave it at that, or to describe utopian solutions that stand little or no chance of acceptance. There is, however, a do-it-yourself procedure for designing trial solutions to problems about technology and people, and this is the art and science of simulation and modelling.[15] If the key features of a situation can be identified (and preferably quantified), they can be put together to form a model, a game, a play, or an opera, whose study may illuminate the real-life situation.[16, 17] A traffic-flow model may indicate the linkage of signal timing; playing bridge may indicate the need for understanding the psychology of one's allies and competitors; watching *Hamlet* may suggest a need for resolution and decision. Models can be built in a rich variety of ways, and their study provides a constructive approach to the better understanding and prosecution of the technological contract.

Modelling may be done by looking down on an entire situation (which risks inadequate attention to vitally important detail) or by assembling detailed pictures of sub-situation (which risks neglect of vital interactions).[11] Experience suggests that good policy requires both of these views, from the 'top down' as well as from the 'bottom up'. Descriptive general terms for these viewpoints have not hitherto been provided. Since there is a recurrent need for such terms, we suggest (and use) *synoptic* to mean 'top down' and *synthetic* to mean 'bottom up'. These are within the definition of the *Shorter Oxford Dictionary*, but both terms have a wide range of other meanings and may be found awkward by some users. If better, more generally understandable terms can be found, we shall want to adopt them. We shall return to the methodology of modelling in Chapter 4.

2. Technology and people

As will emerge in greater detail in later chapters, the unifying factor between the new constraints and problems to be built into the technological contract is the need for a much closer understanding of people and their motivations and emotions. In earlier times, technologists could safely assume that readily definable criteria (speed of vehicles, smoothness of ride, efficacy of antibacterial action) sufficed to define targets and success. Even where 'aesthetics' mattered, these affected only the customer, who bought the garment if she liked it, and otherwise did not. If enough customers, behaving as individuals, reacted favourably, then there was a market.

As soon as the decision to develop and manufacture is influenced by a variety of groups of people—investors when cash is scarce; neighbours when clean air is highly valued; trade unions when wage costs are important; whole populations when aircraft are noisy—the modelling job becomes complex. Then, the technologist totally absorbed in the performance of the plant and the product must lift his eyes from the immediate job, or there will be trouble. The modelling of inanimate systems is, in all conscience, complex enough; working out the balance of cost of materials, conversion, pollution abatement, amenity for employees, and security for bankers, is much worse. For one thing, the rules affecting inanimate systems are stable; those affecting individual people are less so, and those affecting groups are often the least predictable of all. As a start, it is useful to separate the study of people and things, along the lines of Fig. 2.1. There will, of course, be interactions, which can be studied later. One example concerns resources. The price

Fig. 2.1. Relationships between people and things.

of oil since the mid-1950s has been far more strongly affected by the behaviour of people than by cost of winning or any other technical aspect of supply. While the producers were disunited, prices remained nearer the cost (with some tax collected by the producing countries). As soon as the producers came together because of war and politics, the price of oil could be raised dramatically as a political and economic weapon.

The understanding of political influence and preference, as already mentioned, is not usually considered an essential study for technologists. It is often left to sociologists and those trained in history, literature, or the classics, whose studies of politics have often centred on non-technological societies such as the Greek city state,[18] which are very different from those of the democratic, macro-urban, non-slave populations of large technologically based nations. Even those directly responsible for working out rates of payment, incentives, and policies for engagement and retirement of people have usually done little to study the arithmetic of the age-distributions and promotion situations to which current policies may inexorably be leading. An expanding activity tends to recruit younger people and hence have a young age peak, with very good prospects for promotion. When (as is inevitable) rapid growth slackens, average ages creep up, and promotion possibilities diminish. This can set off the selective resignation of the most able, and other processes.

A constructive alliance between sociologists and technologists, as well as being necessary for good decision and management, can also be very good for methodology: each (if he listens) can learn from the other.[19] There are a number of areas for the understanding of people where the technologist's numeracy can help; apart from demography, just mentioned, he can also offer methods for studying simplified conflict and competition and for the design of strategy. With plenty of native wit and without over-elaborate mathematics, the two in concert can predictively treat growth and constraint so as to present a manageable range of policy options for the next year, quinquennium, or decade. Illustrative models of competition can help to avoid investment in the innovations that look attractive but are not powerful or beneficial enough to break in, so as to concentrate on those that are better.[20, 21] Economic growth models exist that can show methods for practically concentrating people so as to develop greater strength. Above all, there are records of success with hierarchies of smaller, linkable, but self-contained models, each covering a homogeneous group of people; these are more manageable than very big models covering large, heterogeneous populations.[22]

Old-fashioned, intuitive guidelines are often based on models, and should not be neglected. Although the new technological contract seems to call for decisions by methods better than 'flying by the seat of the pants', when using sophisticated instruments and data processers for direction-finding and navigation, it never hurts to ask, 'Is the sun where it ought to be?' Sometimes, alas, it is not.

3. Technology and the future

Writing about the future is an ancient art, which has been deployed in many ways.[23] Originally, it most often took the form of religious prophesy designed to draw attention to the likely or conceivable consequences of sin against a particular code or dogma. It was purposeful and effective, because although prophets who were right (and said so) were usually dealt with harshly, many people remembered, felt guilty, and changed course somewhat.[24] Then, and more explicitly, imaginative writers and thinkers sublimated some of their frustration about the stupidity, myopia, and brutality of their contemporaries by writing about utopias, to show how things might be in an idyllic world.[25] Normally, these utopias were either non-political or supposed a very large change in political sophistication. Throughout all this time, seers wandered about in society, usually ending by being indicted for witchcraft, but acting as the centres for mixtures of apprehension and superstition.

More explicit scenarios of technological or geopolitical probabilities are more recent, but it is now a century and a half since de Tocqueville wrote his perceptive analysis showing the likely growth of the power and influence of America and Russia, and a century since Jules Verne wrote a series of engaging fantasies of widely varying technological accuracy. *Twenty thousand leagues under the sea* is probably his best prophesy.[26] H. G. Wells united scientific optimism, political reformism, and an element of doom fantasy in novels that set new standards for futurology;[28] from this sprang the great flowering of science fiction and the much smaller, but equally important, subset of technopolitical writing. These range from John Wyndham's brilliant analyses of the political response to imaginative scientific or technological improbabilities (triffids, chrysalids, krakens, Midwich cuckoos), through John Christopher's disturbingly plausible *Death of Grass* to the hauntingly real *Nineteen-eighty-four* of George Orwell and the earlier *Brave New*

World of Aldous Huxley. The quality of thinking and writing in this sector must rank very high in the long history of serious literature.

The whole of this endeavour over the last two centuries has been almost completely ignored in the conduct of affairs. This has been partly because technology has been conducted projectively and randomly, on the principle of 'letting a thousand flowers bloom', cultivating the pretty ones, and seeing where it all led. The capitalist principle is similar: society, so runs the doctrine, can always find the money for a good invention, to whose number there is no limit. Consequently, the alliance between the two was natural, and was a powerful source of rapid change, much of it evidently beneficial in the diminution of hunger and disease and the increase of *individual* opportunity and amenity. Normative scenario writing: 'This is where present trends could lead in twenty years: do you like the prospect?' was thought to be too academic for governments, treasuries, or businessmen.[29]

As so often happens, the first linkage between the worlds of prediction and reality came in the military sector. Before 1939, new military thinking was only rarely taken seriously by the soldiers themselves, but the very discontinuity of war (and the boredom of peacetime) makes extrapolation seem awkward and scenario-writing attractive. Great military leaders, from Alexander to Napoleon, usually had a synoptic concept of war that was coherent and effective. After Napoleon, Clausewitz[30] provided a base for the thinking of the good minds brought together in the Ecole Polytechnique, and in Prussia also. Maddeningly, those who thought best did less well than the British, who relied on courage, tactical adaptation, the breathing-space offered by island status, and finally the superior competence and great geopolitical relevance of the British Navy to tip the balance. After the very close-run thing of the Second World War, the American armed forces took futurology seriously, and engaged in the most elaborate simulations of missile and other situations, using big computers and assemblies of very able minds.[31] Organizations such as the Rand Corporation came into being.

In private industry and business, there is no doubt that futurology is dangerous if it distracts attention substantially from current competition, operations, and solvency. Big bankruptcies testify to the hazard. And (contrary to the situation in a peacetime army) everyone is busy. At times of money shortage, effort on long-term forecasting can be discarded as a luxury, with no evident penalty. But R and D, inevitably, is working for results no nearer than one or two years ahead, and more of it farther into the future, some decades ahead. Research scientists and technologists have therefore had to choose between hoping for the best, in the expectation that resources would be provided and social problems solved 'on the night', or pioneering the activity of long-range business planning. Usually, they have taken the first course, strengthened by the random nature of invention itself.[32]

The nature of the technological contract, and its increasingly complex constraints, now insistently call on the researcher to engage in indicative business planning—not to say exactly what to do, but to chart shallows, shoals, rocks, and channels.[33] Most industrial research organizations now do this at the level of project choice and prosecution (i.e. at the ultra-micro level), but often with insufficient knowledge of portfolio strategy, company goals, or social and national linkages. This situation now leads to projects that are better than formerly when considered in isolation, but have an unnecessarily high mortality because some of them do not fit with implicit but unspoken company goals, or with unstated but nevertheless real social requirements that can be seen as likely to intensify as manufacture gets under way.

This project mortality is leading some R and D managers and directors to intrude into the field of planning, linking their thinking with that of financial planners about likely future cash resources, and with business planners about the partitioning of these resources between the extension and improvement of existing activities and the opening-out of new opportunities. Because of the remoteness of the time horizon, this is bringing them also into collaborative contact with futurologists, to the accompaniment of grumbling and muttering among operators about conspiracies between long-haired intellecturals. Fig. 2.2 shows the kind of planning hierarchy that emerges, and indicates the tightness of the link that is developing between the scenario-writers and the investors.

Planning exercise	Main considerations
20-year scenario	Long-term opportunities in a changing society
↑↓	
5–10-year plan	Growth/divestment of business areas
↑↓	
1–3-year budget	Cash flow, profits, and dividends

Fig. 2.2. Planning hierarchy.

It is therefore timely that some businessmen are forming entities like the Club of Rome, even though very few of them are yet uniting the resultant findings with business practice. And it is also natural that Herman Kahn's Hudson Institute is making strong business linkages by conducting short-term economic scenario-writing alongside the more radical long-term thinking. In these ways, prophesy is becoming an essential study for the participants in the technological contract, and it is necessary to say, in the pages that follow, how our attitude links with those of others concerned with methodology of planning and future studies.

A number of groups have seen the scope and need for joint modelling of sociology and technology so as to expose the factors underlying growth, constraint, and their interaction. Among the earliest were Jouvenel, Jantsch,[29] and Herman Kahn,[34] who made a major contribution by translating complex and hitherto 'unthinkable' subjects, such as nuclear war, into scenarios about which one could reason. Kahn and his Hudson Institute represent the 'technological optimists' who believe that the adaptive capabilities of technology are sufficient to deal with virtually all of the constraints that arise: societies that can use this adaptive power will grow and prosper, while those that cannot will languish and be eclipsed. Kahn comes near to seeing the postponement of consumption, and the technology it brings, as an end in itself. Confronted with the limited stocks of fossil chemical fuels, he turns to nuclear power and is undismayed by capital cost, economic need for reliability, or technical requirements for safety standards of an unheard-of level: these are problems that technology typically overcomes. Asked about climatic change due to extra technological heat-release and temperature rise, he turns towards means for burying the extra heat, and is generally the technological technologist's prophetic leader: just as trade followed the flag, acceptable sociology will follow technological growth. His followers are scornful of questions about the marginal value of additional technology, saying that there have always been those who said that railways would stop cows from yielding milk, or that there was a limiting speed (around 100 mile/h) beyond which a traveller would suffer loss of consciousness.

At the other extreme are the deeply concerned 'technical pessimists', whose arguments have been reinforced by the imminently disastrous constraints forecast with the aid of the world model constructed by Meadows for the Club of Rome.[11] This group has, as courageously as Kahn, set out to translate scenarios into terms about which one can plan and reason. Unfortunately, the Meadows model appears to be too simple in its treatment of the course of events as natural resources become scarcer: it fails to portray price rise, substitution, and more economical use as plenty recedes, and deals instead in more abrupt 'exhaustion', which has not so far occurred. Inevitably, its treatment of social and population response to resource change has to be arbitrary. For those who do not share Kahn's technological optimism, the Club of Rome's call for population control and resource husbandry, and its questioning of the nature of continuing growth, ring true, and there are strenuous attempts to improve the acceptability of the assumptions within the model.

The second report commissioned by the Club of Rome, entitled *Mankind at the turning point*,[35] maintains the earlier spirit of concern and pessimism about 'orthodox solutions', and proposes nominally practicable solutions which, however, would call for political awareness, sophistication, and unselfishness at a level very much higher than exist at present. In this respect, it lies midway between the standpoint of the earlier utopia-creators and the modern students of social dynamics, for there are as yet no signs that rational argument might persuade people to accept such radical redistribution of world resources. This is not to say that evangelistic means would not succeed.

There are two irrelevancies in prophesy and futurology which we cannot escape merely by trying to arm ourselves with social models and taking up an intermediate position between Kahn and the Club of Rome. Both demand honest, sincere, and continuing self-examination. The first is the age-old and salacious desire to shock people and to elevate and isolate oneself as a superior being who has climbed out of the struggle round the food trough into the temple, thereby acquiring a right to preach and hector, with an enhanced hope of salvation hereafter. Spectacular revelation may be valuable as a shocking means to reformist ends, but it should not become an end in itself; the best prophets identified themselves strongly with those whom they sought to guide and sought to share their sufferings. Forecasters must do the same, and after crying that the end of the world is at hand ought not to retire immediately to eat venison and drink port at an academic high table, doing no more about the people they worry. Secondly, forecasting is a happy hunting ground for computer and software salesman, and it is not very logical to pay more attention to a model because it has been so uncritically and inefficiently built that it requires a big central processer and a lot of core store to run. Complex logic all too readily deflects attention from the scrutiny of unsophisticated and arbitrary assumptions on which the elaborate structure has been built. The good model in this field is usually a fairly simple one, and the good modeller spends more time in worrying about the difficult matter of the assumptions that in enjoying the easy and heady complexity of big computer rooms.

4. Normative technological pragmatism and policy

It was our original thought that the transition from growth to constraint, with which the next chapter is concerned, would be disconcertingly but not intolerably quick. We visualized very large (five to tenfold) swings in the balance between resource prices and manufacturing costs over a period of perhaps ten years—a very short time-interval for technologists to learn and develop a new set of methods. This thought, which generated the plan for the book at the end of 1972, was quickly overwhelmed by events. The sudden arrival of unanimity between oil-producing countries, before and during the Middle East War of October 1973, coincided with the world protein shortage.[36] The economic power of the manufacturing countries, which for so long had divided and ruled the primary producers, was seen to be an unclothed emperor.[37] As a result, producers of other raw materials such as phosphates tried their hand at collaboratively raising prices and found that they succeeded. Expectations of growing consumption by manufacturing workers were clearly threatened by trade imbalance, and willingness to postpone consumption was insufficient to maintain investment. Many other waves rocked the technological boat.

At this point, normal human reaction may be of four kinds. First, one may pursue classical moves to improve one's own position at the expense of others: this method is always represented as the pursuit of social or personal justice by the mover (whether unofficial striker or property dealer) and as iniquity by the sufferers from his action, who resort to more violent response. Such action is obviously more damaging, and therefore more effective in the short term and more disastrous in the long, at times of economic stress. Secondly, one may seek expedients that will mitigate the overall short-term problems, without too much worry about the long-term results. Cutbacks in maintenance, investment, or recruiting fall under this heading. Thirdly, one may identify the long-term policy that would solve the problems but be quite unable to gain acceptance because of a lack of consciousness that proposed present discomforts are really necessary, 'Let's just carry on and maybe the storm will blow over'. Fourthly, having rejected the first two attitudes and failed to gain acceptance for uncomfortable but rational behaviour, one may seek a combination of the least damaging expedients (as a means of getting people to listen) with education in the most effective long-term methods (as an attempt to break down resistance).

The attempt to build social scenarios into the fabric of business planning is a first step along this fourth road. Unfortunately, it is at present a lonely activity; in the public sector, there is no democratic consensus about the options or the preferences. The alternation between right-wing governments (who believe that the technological contractors are simply the technologist and his customer, with the government marginally legislating against extreme abuse of power) and left-wing governments (who do not trust the market mechanism very far, if at all, and want far more interference), precludes any normative inputs to planning. The norms are not shared. Moreover, small- or medium-sized businesses usually lack the resources and awareness that are needed to plan in terms of a new social perception. Consequently, the linkage between scenarios and present action seems unlikely to build up quickly; it needs education, and a great deal of pressure from events.

The position is a difficult one. Academic diagnosis, with action left to others, seems cowardly and pusillanimous. Pessimism, accompanied by a call for draconian reforms to avert disaster, seems unlikely to be heeded and acted upon. Technological optimism seems unjustifiable; the problems seem too large for facile solutions that rely on the continued harnessing of acquisitiveness, accompanied by more material advances. The argument for pushing ahead pragmatically with normative planning, and relying on slow and steady progress, depends on two assumptions; first, that current population control will succeed, and secondly, that there is enough muddled flexibility in the present world political system to provide responses to constraint and pressure that will not provoke disasters such as nuclear wars, widespread famine, or poverty at a level that generates pandemic disease or major social breakdown. The continued operation of the world economic system during 1974 and early 1975 came as a surprise to many; strict logic would have argued that the massive new trade imbalances between manufacturing and resource-owning countries would generate recession through attempts at correction through cutback. In fact, a bewildered combination of the doctrines of Mr. Micawber and Lord Keynes, coupled with a strong feeling that precipitate action, however logical, would probably be harmful, has kept trade and manufacture going. It may be that events will prove that Micawberism has its limits, and that nations spending 5 to 10 per cent more than they earn, as Britain was doing in 1975, cannot carry on by pawning natural resources (like North Sea oil) in the hope that the pawnbroker will feel it inadvisable to keep records or collect the debt. Or perhaps the pawnshop might be burnt down by accident, and the records destroyed.

A very good summary of technological and social hopes, fears, and possibilities is provided by Dennis Gabor's *Innovations: scientific, technological, and social.*[38] He, too, is unable to provide a crisp, neatly packed solution, or to be optimistic about technology alone providing the solutions. His view that the area of social innovation is now more important, and that it carries many responsibilities not yet understood or worked out, is one that runs through the rest of this book and the approaches and methods that it lays out. The explicit building, exploring, and unification of synthetic and synoptic models is a method in which we have confidence: the question is whether it is quick enough.

3

The pattern of technological growth: reliability, economy, and acceptability

1. The nature of technology: the postponement of consumption

Western civilization rests on a belief that society can be progressively improved by postponing consumption and thus setting aside resources for a wide variety of investments.[1] Although it has depended on a variety of political systems, some of these have been especially appropriate for particular kinds of investment. For example, the purchase of military prowess, wherewith to subdue or withstand neighbours, has sometimes been made more efficient by an authoritarian leader, while the more subtle purchase of technological innovation (to cure disease, reduce famine, increase mobility and so forth) seems on present experience to be best done in a capitalist democracy. These are not statements of political preference, for democracies have defended themselves extremely successfully, and it may be that the Chinese communist state has found a non-capitalist means for motivating inventors and innovators.[2]

The belief in the value of postponing consumption has grown hugely during the twentieth century, and an ever-increasing proportion of steadily increasing production has been poured into goods, factories, medicine, education, and science and technology. In Japan in the early 1970s, 35 per cent of the national output was being spent on the long-term capital assets, without including automobiles, consumer durables or clothing: no more than 35 per cent of Western national income is spent on food and fuel, the most immediately consumed items. Table 3.1 shows the U.K. breakdown. A very substantial, and growing, proportion of the population has been employed in the building of productive machines and factories; and in modern capital-intensive industries like petrochemicals, or in research, it is not uncommon for each worker to be backed with plant whose total value is one-half his total lifetime earnings.[3]

Table 3.1. Breakdown of U.K. gross national production 1973

	$£10^9$	Percentage of GNP
Gross national production	63·27	100
Domestic fixed capital formation*	13·87	21·9
Consumers' expenditure	44·86	70·9
of which: Food, drink, and tobacco	13·95	22·0
Fuel, light, and travel	5·74	9·1
Durable goods	4·01	6·3
Housing	6·19	9·8

* This is the figure for gross capital formation. Capital consumption in 1973 (i.e. scrapping old plants, etc.) was $£7·01 \times 10^9$. Hence the National Income was $£56·26 \times 10^9$, and net capital formation at $£6·86 \times 10^9$ was 12·2% of national income.
(Source: *Annual Abstract of Statistics,* H.M.S.O. 1974)

As its simplest, postponed consumption consists of the seed potatoes and seed corn for next year's crop, and the farm animals kept alive for breeding or for draught purposes. At its more rarefied, it consists of space exploration programmes, particle physics, and cosmology (whose relationship to future benefit is not explicit). In between earthy agriculture and esoteric science and art lies technology, occupying the extensive middle ground, and employing a very highly educated tranche of perhaps 2 per cent of the population, backed by many others. This group has been accustomed to steady growth, unquestioning support from society, periodic bursts of wild applause, and very little criticism, either constructive or hostile. It has therefore got on with the job as seen, without too much attention to underlying principle or the possible emergence of criticism or major constraint. To be sure, technologists have not usually commanded all the money they would have liked, but in recent years they have spent sums that, forty years earlier, would have been unthinkable. The tasks of making vehicles bigger or faster, or television sets thinner or less unreliable, were justified impatiently, if a question was asked, by the use of the nebulous word 'progress'.

Technologists are thus poorly equipped to meet hard constraint, whether from intellectual doubts about the value of some kinds of progress, from concern about the effect of urbanized economic growth on the environment, from increasing difficulty in finding and winning resources (particularly energy) or, worst of all, from a basic but poorly articulated and understood decision by society that such massive postponement of consumption might, from now onwards, gradually cease to be so necessary or beneficial.

In a way, it is remarkable that scientists and technologists tend to regard constraint as unnatural and unfair, for, since resources are limited and desire is infinite, it follows that constraint must be usual and complete liberty atypical. It just so happens that a social group occasionally impresses its peers sufficiently to be given special treatment. Armies have sometimes been privileged, and technologists enjoyed a considerable freedom for two or three decades, conspicuously helped by their great success in crucial wartime tasks.[4] A society that had just escaped from primitive concern with needs—the thrusting back of famine and disease—was able to afford special resources for the pursuit of opportunity, and to give special licence to the people who had contributed substantially to the improvement of food supplies, social conditions, and medicine, that had reduced disease and hunger alike. Inevitably, these processes then generated demands on resources, which were further amplified by inexorable population growth. Having generated the possibility of new materials and devices of all kinds, technologists have found their growing organizations rapidly checked by the consequences of their very success.[5] And, having enjoyed no more than a single generation of the simplicity of socially encouraged growth, their talents have now to be directed to the much more challenging complexities of the balance of growth and constraint.

At present, therefore, the technologist is feeling badly let down, and he often reflects that society is capricious and disloyal to all those who give unstinting service and thereby work themselves out of a job:

> God and the doctor we alike adore
> But only when in danger, not before;
> The danger o'er, both are alike requited,
> God is forgotten, and the Doctor slighted.[6]

2. General difficulties in maintaining technological advance

Any technological development that has gone through a period of relatively unconstrained growth inevitably faces the prospect of disarray. There is nothing new about such a pattern, and many technologies that are now deeply embedded in our society have shown it. Coal-mining, steam engines, textile manufacture, shipbuilding, steelmaking, and railway construction all hit constraints after the initial honeymoon period.[7] The pattern of the difficulty was different in each case, but the response was the same: the industry could not longer support the invention and enthusiastic effort that had brought it into being and carried it to prosperity, and the dynamism was lost. Safe men pursued safe policies of retrenchment, and were followed by dull men lacking the wit or ability to do otherwise than carry on unenterprisingly. These dull men were often the sons or grandsons of the original entrepreneurs, educated for and accustomed to a level of affluence that tended to deprive the struggling business of such money as might have been devoted to a search for a better way ahead. Indeed, the new directors were often actively hostile to change to any kind, having established conferences and agreements of a defensive nature. Nemesis frequently took the form of an inability to pay the rising wages or provide the better working conditions called for by general social advance. Inefficiency thus first called for government subsidies and then (as these were misused by poor management) for nationalization.[8]

The plain truth is that technological advance requires imaginative leadership. Only then will it generate products commanding prices high enough to pay and motivate a vigorous workforce that feels it is getting a fair deal by current standards. A well paid but badly led and motivated work force will not do, nor will a highly subsidized but uncommitted and dull management. A combination of the two, which has sometimes been achieved, is quite disastrous.

Nemesis also had other manifestations. One such was simple over-optimism which caused investors in different but competing businesses to assume that they could all obtain a large share of an overestimated market. Private coal mining was unable to pay an acceptable wage because it failed to recognize the need to concentrate on thicker seams which could finance better machinery and methods. Railways provided for even more traffic than a burgeoning economy could generate, and commitment to high fixed costs then made the new network unprofitable. Two systems linking London to Aberdeen, to Exeter, or to Norwich; a station in every village; a chaotic goods system involving a nightmare complexity of marshalling yards; all these arose within fifty years.[9]

Another catastrophe could be caused by the territorial transfer of the simpler forms of a technology to a developing country. Thus when cotton spinning and weaving became widespread in the lower wage-cost country of India it acted to constrain the British industry. The remedy of tariffs and quotas against imported yarn or grey cloth failed to motivate further U.K. technological advance, and in any case conflicted with the free trade, cheap food policy that was an integral part of the entrepreneurial liberal ethos, founded on the Doctrine of Comparative Advantage. This theory had attracted the vigorous founders of an industry but repelled their offspring, who saw it as the enemy of comfortable protected 'arrangements'. Yet a third route to disarray was the simple failure to continue innovating. The attachment of first-generation innovators to a new conservatism was usually based on an inadequate realization of the further steps that were possible and beneficial. It is hard to accept that a shining new very high pressure technology, with large sums of money invested in it, must inevitably give way to a catalyst invention that permits better or more adaptable product mixes to be made more cheaply at low pressures. Because of such disarrays the Germans captured the British-born dyestuffs industry, and the Japanese have captured a whole series of European industries.

As any technological industry grows it therefore needs increasing competence and determination if disarray is to be avoided. The pattern of skills needed is continually altering: a lone inventor needs a small team of developers, then a big team of industrialists, and then the continuing efforts of a very competent team of managers and innovators to prevent the juggernaut from sinking into the mire. Without such a succession, the industry is in danger of being immobilized by a discontented Luddite mob created by managerial malpractices.

3. The background to the present phase

Until about 1950, the growth and constraint of the individual industries mentioned above had been separate in space and in time; thus any stress could be partly taken up by the societies in which they were embedded. Railways in Britain suffered their traumas intermittently from 1860 onwards; coal mining and the docks began their troubles around the turn of the century, cotton textiles 10 to 20 years later, and the woollen industries around the middle of the present century. Superannuated railway engineers found their way into other forms of engineering and founded industries (e.g. automobiles) that in due course employed people displaced from textiles and shipbuilding.

Now, however, technology and society are much more closely intertwined. The impact of constraint on growth is international and simultaneous, rather than local and intermittent. The unifying factor has been the integration of science and technology with the educational system and the economy at large. Such interdependence has been of great net benefit, but as a result problems that start in one industry can readily spread into others. This makes it unsafe to ignore remote or general questions simply on the short-sighted grounds that one's own firm or sector of society enjoys good health and is well managed.

In 1970 there were not far short of half a million British technologists; in 1900, no more than a few thousands. Yet nearly all the devices of the twentieth century had been basically invented before 1914: powered flight, telecommunications, road and rail transport, systematic building, surgery and chemotherapy, and even the computer. For the most part, these had come on a 'craft' basis, by the application of trial and error to the making of useful inventions from a modest range of materials. Why, then, has society seen fit to create such a large population of technologists *after* the seminal events on which their trade is built?

The rather curious answer is that technology is concerned with the underpinning and dissemination of basic invention, and rarely with invention itself. Invention is artistic and difficult to plan,[10, 11] while technology depends essentially on good planning and the extension of known techniques. The three purposes of technology are to make inventions reliable, economic, and acceptable. To these purposes the application of science and logic has of course been crucially helpful; it is difficult to assert that they have been essential, for many devices have been made extremely reliable by craft methods alone. The instruments of the orchestra are a prime example of this, for the ear is a sensitive dectector of inconsistencies of tone. But even if a 500-seat aircraft could have been developed by empirical craft methods, it would assuredly have been at the cost of far more time, resources, and deaths than by the scientific approach. Science and logic have a life of their own in a world of their own, and have valid intellectual objectives of their own, quite apart from technology. But scientists have greatly extended the resources made available for their own purely scientific endeavours, as in cosmology or biology, by the demonstrable fruitfulness of their generous and willing alliance with technology.

During the past fifty years, one of the technologist's most formidable difficulties has been to keep pace with changes in the relative importance of reliability, economy, and acceptability, as his wares have increasingly penetrated everyday existence.

To begin with, reliability dominates: if the device could be made to work, it would usually prove cheap enough to develop in some context or other. Radio was initially an expensive invention that was found worth while for marine or military communication, and polythene an expensive material without which high-frequency electronics were daunting. A special introductory use has normally led first to lower costs and then to economics good enough for mass use: thus radio and polythene are both now found in the kitchen. At this point, the technologist is forced to elevate economic requirements to equality with reliability. To be sure, he always had to concern himself with cost, but in such a way as to treat it secondly and separately. Now, however, he must recognize that such and such an object, however well made or attractive, cannot possibly sell at prices above some stated value, unless of course it can be given an artificial scarcity value so that it becomes a symbol of wealth.[12]

With the widespread dissemination of technological devices, acceptability now has to be similarly elevated.[13] One man's midnight take-off can be a community's insomnia, and although technologists have always been concerned to clean up the effluent of the process they are developing it has been a tertiary rather than a primary matter. The seriousness of side-effects, initially very small, grows more steeply than the benefits of most technological advances; it is now commonplace that environmental requirements have to be rated on the same basis as reliability and economics.

Clearly, any society in which the majority of families own a motor car, television and radio sets, and a variety of equipment consuming up to a score or so kilowatts of gas or electricity is much more interdependent than when the basis consists of bicycles, coal fires, hand sewing machines, and oil lamps. This interdependence makes very heavy demands on good human relationships and good management of nations, towns, and networks of many kinds. Technology which requires a degree of social cooperation that we cannot or will not provide is quite simply unacceptable. The technologist therefore cannot now relegate any of his major considerations to a secondary or tertiary position: all three, reliability, economy, and acceptability, are equally important.

4. An historical example: the steam engine

It is instructive to look farther back at the less spectacular way in which economics and then acceptability have come into industries that are now mature, but where the founders were able to concentrate initially on reliability. What one sees is great waste and unnecessary delay; in retrospect, the able technologists, had they been able to take a synoptic view earlier on, could have done a great deal to adapt their own skills to the anticipation of future needs. Instead, after the early surge of activity, there was usually an economic overshoot and a depression that dispersed the most able people. The new needs were not attended to, and stagnation continued for an unnecessary period. It may be that the capitalist machine, which is so good for the initial surge, cannot operate satisfactorily for the transition—Keynes, Roosevelt, and Galbraith notwithstanding. Whether this is so or not, we are certainly dealing with matters in which politics play a vital part.

With the foreshortening which always accompanies the historical view, we tend to think of the advent of steam power as a rapid succession of Cornish mine pumps, James Watt, George Stephenson, the railways, and the mills. The reality is very far from this.[14] It was in 1700 that Savery first formulated and patented a steam pump as the 'Miner's Friend', but 1711 before Newcomen, assisted by Calley, who was a good plumber and therefore able to make things work, introduced the first pumping engine at Dudley. Coal was vitally needed because of the energy and resources crisis arising from the felling of forestry reserves and from the consequent demise of the charcoal industry. Hence these initial machines of adequate reliability (but with a thermal efficiency of about 0·5 per cent) were installed for over fifty years with relatively little economic improvement. This period was about twice as long as the lifetime of two of our own most stable modern pieces of technology—the Ford Model T (1911—27) and the Volkswagen Beetle (1938—   ), and longer even than the life of the Lancashire boiler.

It was not until 1767 that Smeaton began a systematic study, on good value-analysis lines, of the factors affecting efficiency, and actually measured the amount of water pumped per pound of coal burnt. He was able to double the figure of 6×10^6 ft lb work per pound of coal, by improvements in cylinder-boring technique that enabled him to make much bigger cylinders at Carron; this took him about seven years. With this piece of optimization (to 1·4 per cent efficiency), the economics were satisfactory for more than another half century, partly because the machines burnt low-grade coal that was not worth the very high cost of transport to markets that would pay for it. The last engine of this vintage came out of service at Park Gate in Yorkshire in 1934, after operating for a century with virtually no attention.

At almost the same time as Smeaton's optimization programme, Watt embarked on the innovations that made the steam engine into a prime mover as well as a pump. His partnership with Boulton was crucial both for production and for a patent deal that extended his monopoly for ten years. Watt's separate condenser and governor-operated controls improved the thermal efficiency another threefold, and enabled him to sell about 500 machines, about 300 of which were for driving machinery. His engines, best for large mills, paved the way for the higher pressure Cornish machines whose compactness made them more suitable for smaller mines and water works, and for vehicles. But by now this burst of attention to performance was crucially dependent on economic factors, and it is especially noteworthy that George Stephenson's work on steam locomotives started in 1814 partly in response to the rising cost of horse fodder!

Although the Victorians were more responsive to economic progress than environmental acceptability, railways were soon regarded as dirty. In the 1860s coke was being used instead of coal because of legislation compelling railway engines effectively to consume their own smoke. But the collapse of the first railway boom had deprived the railways of the people whose efforts could have paid the broad attention to economics and acceptability that would perhaps have prevented the decline of the railways during the rise of the motor car. One who sized the problem up imaginatively was Brunel. He saw the economic and environmental advantage of building a big stationary prime mover and transmitting the power; his own approach, at the end of his life (1847), was his pneumatic railway in Devonshire. Cable traction was briefly tried in the Glasgow underground. Only in London did the urban environmental imperatives call forth a team of people able enough to seize rapidly on electrification of an entire urban transport system. It is true that elsewhere in Britain there were good locomotive designers who operated well within the confines of their trade; but no one recognized the true nature of the opportunity as did the French, Swiss, and Dutch, who electrified rapidly and as a result had cleaner, cheaper, and more reliable trains, moved by rotary rather than reciprocating machines.

5. A modern example: the chemical and related engineering industries

In the twentieth century our background is one of industrial specilization far more intensive than ever before. This changed situation must be understood if we are to develop adequate social responses to modern constraints.

A whole series of industries has been concerned with major enabling inventions that have first made the late nineteenth century inventions reliable, and then so cheap as to penetrate very deeply into the social structure. There is little general appreciation of the depth of this penetration. Some simple questions drive this home: first, how many electric motors and switches does the ordinary Western family, living on the average national income, possess and expect to work? Typical answers would be 10 and 50 respectively.

Secondly, what is the annual per capita consumption of steel and the more modern materials? The answers, 400–800 kg of steel, 40–80 kg of plastics, and 20–40 kg of textile fibres (one-third as garments, one-third as domestic furnishings, and one-third in tyres and industrial uses) make it possible to visualize the enormous amount of industrial activity that is taken for granted and whose sudden withdrawal would cause grave bewilderment and disorder.

The industries concerned—motors and petroleum, electrical, food and home consumables, textiles, and chemicals—were often producing speciality items in the 1920s and 30s. They were still pursuing reliability and performance: their goods were costly because of small scale of production, and they were bought mainly by the middle and higher income groups. During the period 1920–50, a series of major enabling inventions, new technological economies, and greater real wealth, brought motor cars, radios, television, telephones, a variety of domestic amenities, prepared foods, attractive garments, and new medical aids within reach of a much higher proportion of the population, if not yet an actual majority. Affluent specialities had become commodities. Economic behaviour was moving away from the sharing of insufficient goods against a background of insatiable need, to the distribution of amenity goods against a background of apparently insatiable opportunity. But growth could still be assisted and stimulated by further cheapening, and it was dangerous not to respond to this pressure. Failure to do so gave competitors greater economies of scale, and Fig. 3.1 shows some price movements within this period that would have left a constant cost producer completely stranded. The technologist now had to pay equal attention to price (or cost) and performance: neither was sufficient.

By about 1960, the new materials (fibres and plastics, for example) were competing successfully with the well-established products (cotton, wool, wood, and metals) even when these fell to what seemed to be depressed price levels. Consequently, what seemed to lie ahead was a future of growth from three causes: capture of market share from traditional products, population increase, and increased *per capita* consumption. The gap between the affluence of developed and developing nations might be widening, but both were insistently demanding

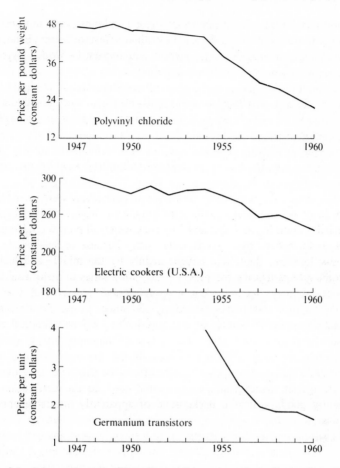

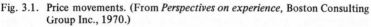

Fig. 3.1. Price movements. (From *Perspectives on experience,* Boston Consulting Group Inc., 1970.)

the new products, and volume growth of 15—20 per cent per annum was not unusual.

The appearance of constraints was evident to the technologist before it was to the market forecaster. A number of problems arose in factories while the markets were still expanding at maximum rate. The massive growth in the output of goods might have been accompanied by an equally massive but unacceptable growth in effluents. Strenuous efforts were needed to live within the dispersal capabilities of rivers and air-movement. In some cases failure to do so led to great unpleasantness, and technologists in Los Angeles were the first to be concerned with

29

the new varieties of smog.[15] Heavy metal effluent levels needed to come down substantially, for toxic materials such as mercury and lead were now being used in significantly large quantities as seed dressings, carriers in production processes (e.g. chlor-alkali cells), and fuel additives. In addition, specialized production accompanied by low levels of manning meant that very small groups of people could bring whole industries to a standstill, by strikes or other industrial action. Design for acceptability moved further and further up in importance, and nearer to the start of innovative developments.

This concern is now being expressed even more overtly in some countries through the formalism of Technology Assessment, in which the secondary and even tertiary effects of any prospective innovation are subjected to searching analysis before introduction. We have now reached a position in which it is inconceivable that any major new technology could make progress without early and broad consideration of acceptability, economics, and reliability. Unless the systems for training technologists take this fully into account we shall be misusing our resources and harming our society by default.

6. Educating the technological community: the 1970 transition

From the time, around 1850, when scientific ideas were first constructively brought to bear on the crafts, until the late 1960s the growing number of European scientists and technologists formed a community whose pattern of education and employment was mainly specialized and separate from that of the non-scientists. The consequences of this have been explored and discussed extensively: Lord Snow's concept of 'two cultures' may have been slightly exaggerated, but it drew attention to an important social division.[16] Essentially, the spread of applied science created opportunities for employment faster than new scientists and technologists qualified, and it therefore appeared wasteful for anyone with scientific qualifications to work otherwise than as a specialist practitioner. Economic forces created an expanding hierarchy of well-paid and secure jobs, particularly after the 1939–45 war. The expansion of scientific education to fill the gaps was a natural consequence. This expanded education was mainly concerned with the fundamentals of science and with capabilities relevant to the performance and reliability of technological materials, machines, and systems. Engineers learned a little economics, but scientists, for the most part, did not.[17]

A number of signposts showed that there was considerable scope for employing scientific skills outside science and technology themselves.

One was the wartime introduction to military and naval strategies of scientific methods that conflicted with conventional professional theory, but were evidently more effective.[18] Experimental analysis based on good statistical techniques yielded better methods for convoy protection, submarine hunting and gunfire direction, using given resources. This field of 'operations research' was opened up by very able university scientists urgently seeking to make a contribution to nattional objectives, under circumstances more demanding than those of peacetime, and with licence inconceivable under normal, well-ordered conditions. The successful operations researchers included nuclear physicists, biologists, and chemists, who were using the essentials of experimental method that they normally deployed on totally different laboratory problems.

Another signpost was the practice in big science-based industries of moving able men, usually in their thirties, out of specialist science and engineering into general management. Electrical, engineering, textile, chemical and oil companies have all been successfully and profitably run by erstwhile specialist scientists. An even clearer signpost was the recognition in the U.S.A., that formal education could be given to help engineers and others to apply their skills to business and administration.

During the 1960s there was increasing pressure to employ some scientists outside science, simply to improve the general running of society by using standard technological practices in urban and national design. In addition, operations research techniques were brought back after a period of post-war neglect into road and rail transport problems, airport traffic handling, and many other fields.[19] Even within scientific industry, there turned out to be plenty of opportunity to improve the scheduling of production, operation, and distribution. But tradition still led most of the best scientists into the practice of the skills in which they had been trained, and left the top management of government services, even in technical areas such as posts and telecommunications, or aviation and transport, to non-technologists. Misunderstandings between administrators made nervous by their lack of technical comprehension and scientists made arrogant by being denied authority increased the ill effects of the lost opportunities.[20] Only in the private sector was there anything approaching a reasonable synthesis of skills, and even there inefficiencies occurred because of the paucity of the able humanists and sociologists to be found in this sector.

Against this background, the supply of newly graduated scientists and engineers exceeded the number of places for specialist practice in most European countries from about 1969. At about the same time, military and space research cutbacks produced the same result in the U.S.A. Since educational trends take a long time to generate, and then continue along the same lines unresponsively when extra places and facilities have been provided, the first reaction in many places has been to conclude that scientists were now being 'overproduced' since 'needs' had been more than satisfied. The alternative view—put urgently as a counter-proposition by those who, in the 1960s, had been urging the wider employment of scientists in 'affairs'—was that at long last there were enough scientists to begin to pursue the opportunities for integrating technology into its social environment. The best scientists were becoming available to apply their methods and skills in helping run chancelleries, embassies, town halls, and small businesses.

Two difficulties immediately become apparent. First, scientific education had become very closely linked with the specialist world,[21] so that unusual courage was needed by a science graduate wishing to deploy his skills outside science. He was met by selection systems accustomed to dealing with able and fluent candidates who knew history or economics, and who knew that real-life decisions could not be reduced to the high-probability propositions generated in the laboratory. Moreover, specialists were still required in fairly large numbers, so that the new need was not wholly substitutive. Secondly, in some subjects, notably chemistry, school leavers were deciding, rightly or wrongly that the specialist university courses were dull and less relevant than in other subjects where a broader view could be taken. This trend had started in the heyday of specialist scientific employment, and now was reinforced by the feeling that a science or engineering degree was no longer a safe route to a secure job of known character. Consequently, the opportunity to employ scientists and humanists in combination, with a new and powerful synthesis of skills, was threatened. In some faculties, much less able science entrants were admitted; these students unfortunately were those least likely to develop broad capabilities. In others, numbers fell. Table 3.2 shows typical instances.

We have already said that there is a new equality between the needs to make technology reliable, economic, and acceptable. Hence the new employment imperatives have arrived opportunely, but a heavy and difficult task falls on the shoulders of the present generations of new graduates, their teachers, and their employers. It may even be that the relationship between the technologist and his community has been changing at a rate that is too great for the educational system. It is

Table 3.2. Admissions of full-time students to U.K. universities

	Chemistry	Physics	Chemical engineering	Mechanical engineering
1967	3308	2315	869	2027
1968	3035	2189	900	1984
1969	2960	2293	893	1914
1970	2889	2450	956	1847
1971	2723	2509	901	1953
1972	2529	2385	743	1765
1973	2107	2093	622	1678
Percentage fall 1967–73	36	10	28	17

(Source: Universities Central Council on Admissions)

difficult and demanding enough merely to keep modernizing the professional and technical components of the science and engineering syllabus as theories, materials, and methods of calculation and treatment all change. Having arranged over the past fifteen years to teach the new subjects of organometallic chemistry (highly relevant to catalysis and biology), polymer physics and chemistry (important in materials choice and again in biology), group theory and other theoretical studies to underpin spectroscopy, and so forth, a chemistry professor can only groan inwardly (or outwardly) when he is told that at least half of his pupils are now being educated *through* chemistry rather than *for* chemistry. Consequently, his informant adds, they will need to know economics, psychology, behavioural science, some history, a good deal of geography, and some languages. This may well be better dealt with by the technologist himself as an adult learning problem, rather than by the teacher as a formal teaching problem. The new needs, which by no means abolish or supersede the need for specialism, especially in new fields of activity, generally require interactive skills and the ability to form, influence, and motivate cooperative groups of colleagues. They further call for perceptive interpretation of the social aspirations of large numbers of people, either as customers or as citizens in need of social services. We shall consider the definition and provision of new patterns of skills in Chapter 9.

7. The need for future growth, and the use of 'artificial emergency'

It is periodically suggested that technological advance itself may have overshot some desirable or feasible level. As long as it was concerned with the abolition of famine, the provision of clothing and shelter, or the reduction of disease in a moderate world population, there seemed to be no cogent argument against it. Once it created an increasing volume of work for an expanding population, whose age-span was lengthening because of success in attaining the basic objectives, then overshoot in terms of the arrival of overcrowding and aggression, loss of amenity, and severe pressure on resources became probable if not inevitable. We shall be examining in subsequent chapters the point at which overshoot arrives. But some able economists and politicians are now again questioning the need for economic growth, while others are insistent that society is dependent on growth, and cannot escape from it. Should we, perhaps, close the argument at this stage and simply say, 'enough is enough. Growth has gone far enough and must stop'?

It would seem that the protagonists and antagonists of growth are thinking on different time-scales: the former about short-term political motives, and the latter about longer-term physical and psychological limitations. It is possible to reconcile the two views and to synthesize a scenario that is internally consistent. This requires two basic reforms: one political and one biological.

The first is to dissociate status from wealth and the ownership of goods that require the massive use of resources and energy, and the substitution of the notion, already partly established, that status has different meanings for different people. The rationale underlying such a notion is that once bodily needs have been adequately met then the only true wealth is that of the mind. The provision of opportunities for flaunting such wealth needs careful design. It may be that a 'wealthy' member of society will come to be one who can exercise a high degree of personal choice and yet still consume resources in only moderate amounts. Thus, status, as newly defined, might even embrace such items as the climbing of mountains, the painting of pictures, attendance at (or participation in) sporting events, the writing or study of poems or novels, or the composition, performance or study of music, opera, drama, or ballet (classical, pop, folk, or jazz).

The second, biological requirement is, of course, the control of population, whose exponential growth clearly cannot go on for ever. But given population control, the limitation to growth in welfare becomes the limitation of human energies and the human mind, on the picture just presented. Resource-based goods such as motor-cars, houses, machinery, clothes, and food then become services to be provided on an adequate basis, much as main drainage, water, refuse collection, roads, medicines, and education are now furnished. Arrangements would need to be made either for manufacturing activity to be its own reward as in the craft era, or for it to be acknowledged by medals, titles, special seats at sporting or cultural events, publication of works with special splendour and acclaim or other acceptable and conventionally esteemed baubles that were not heavily resource dependent.

But why, it must be asked, do we need a change of concept; why such a fundamental shift to a different scale of measurement to enable us to go on growing? Why not take refuge in doctrines of the equality and brotherhood of man, saying that once we have reached a gross world product of (say) $3000 (in 1974 values) per head, we will all undertake to be members one of another and leave it at that? The answer lies in the need to avoid a social structure that fixes everyone's position and pattern of activity. For man is eternally competitive. If every advance toward perfection has to be accompanied by someone else's equal and opposite displacement from perfection, then advance toward beatitude will be resisted by those who are, in consequence, cast out and identified with the cast of Brueghel's *La Chute des Anges.*

The idea that everybody's competitive urges can be satisfied, at least in part, is the basis on which Western economic democracy has so far been conducted. By sharing economic growth in favour of those who are less privileged, or who try particularly hard, it is possible to unite haves and have-nots in a common endeavour. The haves are not asked to sacrifice any of the amenities they currently enjoy, but are told that they must continue to do well and remain credible in order to retain them; the have-nots are told that higher productivity (generated by the technologist) will bring greater personal opportunity in a fairer society. All can win, and all can receive prizes.

The trouble comes when economic growth ceases for one reason or another. The have-nots then expect continuing material advance, but the haves are not prepared to accept actual improverishment cheerfully. They reflect that although they may still have pleasant places in which to live, work, and be educated, they have already been deprived of a great deal of personal service, and it is asking too much to expect them to give up a motor car or foreign travel as well. Further, the aspirations of the have-nots (many of whom, by this time, are by no means poor) are too great to be met simply by soaking the rich. Since what is being requested is money, governments print more; and since the excess is backed by no more goods, there is inflation. This over-simple view might seem to suggest that inflation can occur only when real growth ceases, which is demonstrably not so; we shall return to this topic in more detail in the chapter on capital and finance. But aspirations could be met by esteemed physical achievement (such as athletics, climbing, or swimming), competitive achievement in games, whether physical (such as football), or intellectual (such as chess), artistic achievement (individual or collective musical performance, painting, writing), social achievement (successful welfare work; conversation), or craft achievement (carpentry, gardening, do-it-yourself). Wealth could be seen partly as esteemed performance, and partly as the opportunity to watch, comment, criticize, and discuss; this wealth could grow without limit. Its generation and consumption could initially be derided as circus-building, but it would cause fewer coronary thromboses than overeating bread.

For several decades yet, however, society must be provided with more resource-based wealth. This is because further population rises are unavoidable and because novel growth concepts striking so deeply will inevitably take time to be developed. The technologist must therefore join in the provision of the old growth as well as the invention of the new. It is in this connection that emergency conditions such as those of the present hyper-inflation, suggest a search for emergency solutions. One such is the recapture of 'wartime spirit': sacrifices were accepted in Britain in 1940, in Holland (to rebuild the port of Rotterdam before rebuilding houses) in 1946–9 and in Germany and Japan during the postwar years. More recently, the U.S.A. devoted a significant part of its economic growth to an emergency plan to land men on the moon, to boost national morale. Is it true that technology *generally* is advanced by wartime need? If so, can such benefits be attained in peacetime, by the proclamation of a 'synthetic emergency'?

Analysis of true national emergencies, and of the two world wars in particular, casts considerable doubt on the idea that emergencies produce general technological benefit. Fig. 3.2 and 3.3 show a series of economic indices plotted for the U.K. across the Second World War;[4] only agriculture and air travel seem to have benefited notably. In both of these cases, deliberate steps were taken in wartime that peacetime conservatism would have inhibited: the provision of substantial agricultural subsidies to promote national food production (in the face of the submarine blockade), and the payment of the large entrance fee for the aviation turbine (for faster military aircraft). In both cases, the necessary underlying science and invention had come in peacetime; what was needed was a political boost for measures already known to be practicable.

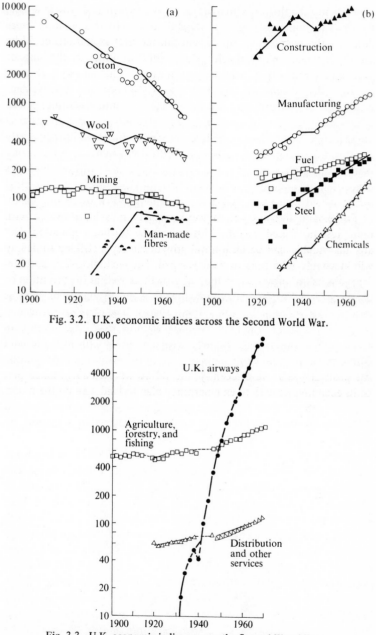

Fig. 3.2. U.K. economic indices across the Second World War.

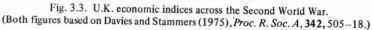

Fig. 3.3. U.K. economic indices across the Second World War.
(Both figures based on Davies and Stammers (1975), *Proc. R. Soc. A*, **342**, 505–18.)

Similar phenomena permitted operations research to arrive in wartime (as already noted), and the introduction of methods for collection of income tax by rough-and-ready wage deductions week by week. Both would have been fought tooth and nail in peacetime as thin ends of dangerous wedges. Thus, what national emergencies seem to do is to remove political obstacles from the pathway of technological developments that are already delineated and are of sufficient benefit to justify some iconoclasm. Much of the wartime technology so accelerated is of no peacetime use, and spin-off is not usually of major importance; and many peacetime R and D programmes were severely retarded. It can therefore be argued that a 'synthetic emergency' can be used to displace technological conservatism but not to stimulate actual invention or long-term science. Obstacle clearing is very useful, nevertheless, and it may well be that medicine may soon be in need for a boost of this kind. It is certainly a technqiue that technologists should seek to understand, but by itself it is clearly not the way to generate future economic growth.

The modern technologist is left very clearly with the need to discover what a changing society requires from him, and to shape his own education and contribution accordingly. But in such a complex area, in which conflicting messages abound, he must first be sure of his ability to communicate those perceptions he wishes to share with his colleagues and customers. Before examining the constraints and responses in more detail, therefore we look in Chapter 4 at the whole concept of models and modelling.

4

Models and their use by technologists

1. Concepts and cartoons

The whole of physical science, and an increasing proportion of biology, is based on the mathematical correlation of numbers that describe the state and behaviour of real objects. For example, telescopic observations on the positions of planets were assembled into Kepler's empirical laws. Then these laws were shown to result from Newton's laws of motion, with the addition of a concept of gravitational attraction between massive objects, attenuated by the inverse square of the distance between their centres. From this point onwards, the whole subject of mechanics was built on the solid foundations of simple concepts that could be used to model physical systems, ranging from interplanetary rockets to primitively conceived atomic systems such as the Rutherford–Bohr atom. As Sir Peter Medawar has observed, 'In all sciences we are being progressively relieved of the burden of singular instances, the tyranny of the particular. We need no longer record the fall of every apple.'

Such mathematical correlations are in effect models. But models are not confined merely to numerically precise scientific laws. The chemist is perhaps best equipped of all physical scientists to comprehend the breadth that modelling offers, because his subject varies so greatly in complexity. As a practical matter, he can sometimes do best with numerical arguments, sometimes with conceptual arguments, and sometimes with semi-empirical compromise. It is possible to embark on a complete wave-mechanical treatment of the four components of the hydrogen molecule (i.e. two protons and two electrons) and emerge, perspiring gently, with good predictions of spectroscopic behaviour.[1] To do the same for the hundreds of particles in UF_6 or penicillin is possible in principle, but almost certainly not rewarding in proportion to the effort. Most chemists would agree that, for the time being at any rate, it is sufficient to use wave-mechanical output to generate concepts.

These concepts can then be used as a basis for a model of atomic and molecular structure and behaviour that permits the semi-quantitative interpretation of an impressive array of observations, from the details of spectra to the descriptive orderliness of the Periodic Table.[2] Almost as rich a yield of valuable ideas has come from the simple yet elegant models of the shared-electron covalent bond as formulated by G. N. Lewis,[3] or the tetrahedral carbon atom[4] and symmetrical benzene ring.

Flushed with two centuries of success with concepts and models of this kind, what could be more natural than that physical scientists should try applying their skills to human or sociological systems such as economic networks? These scientific and technological modellers had the strength of proved success in their own time-invariant fields, with a less physically diseased, more broadly enlightened and aware, and more affluent society as one of the main results of their work.

Their modelling techniques had not yet been applied to the study of relationships between unemployment, economic growth, and productivity.[5] A major handicap is a strong feeling that there should be 'economic laws' as simple and durable as those of the physical world. In fact, a changing society almost certainly changes its interactive pattern very fast. In 1931, unemployment brought real hardship to the individual and led to hunger marches. In 1971, it led instead to strikes against plant closures, since unemployment, while unwelcome, did not quickly undermine all of a man's family aspirations. Consequently, any 1971 model with 'unemployment' as an independent variable must be different in structure from a 1931 model. For better or worse, there are no Newton's Laws of Unemployment; the correlations, such as they are, must be ephemeral.

The technologist, therefore, for all the strength of the scientific paradigm behind him, must adopt a different approach to human assemblies. If the underlying correlations are ephemeral, he must carry out his work fast enough to propose actions soon enough for them not to have been overtaken by changes in the characteristics of the system. Often, therefore, he is obliged to undertake an essentially artistic approach in which he seeks to determine the really important variables, to which the outputs are most sensitive. In this way, he can model quickly and suggest policy options in time for them to be applicable and useful.

In many ways the skills demanded of a technologist as soon as he ventures outside the realm of purely physical laws closely resemble those of the cartoonist. Over many years a distinguished band of cartoonists has concentrated fact into striking sketches; some elaborate, and some starkly simple, like those of David Low.[6] The student of politics is guided from the 1930s into the 1950s by the behaviour of Colonel Blimp and the Trade Union carthorse, much as the chemistry student is guided through the natural products by the behaviour of aliphatic and aromatic carbon assemblies. But, it must be stressed, it is the misfortune of students of politics and economics that their basic relationships are constantly changing, and may become quite different within a lifetime. It is impossible to illuminate the behaviour of Tory politicians in 1975 by pictures of Colonel Blimp in his Tuikish bath, or the frenetic behaviour of militant workers by drawings of a well-understood, stubborn but useful draught animal. Indeed, we are beginning to realize that David Low's period was one of conventional political stability that was a gift to the cartoonist; at present the conventions of 1926 have vanished, so that even such basic phenomena as 'democracy' or 'the law' are seen as being dependent on fragile consensus rather than absolute principle. The chemist has built much of his subject on concepts that he has had to recognize as fragile, and as quite likely to be replaced by something better. This, it may be hoped, has prepared him will for the special need for humility, judgement, and speed in social and economic modelling. His scientific paradigms are nearer than those of the physicist to the realities of economic behaviour.

2. What is a model?

A model is more than a concept used to codify or exemplify scientific law, and more than a cartoon drawn to highlight human actions and responses. We shall show later in this chapter that models are in effect a new kind of language. Most of the *Oxford Dictionary's* definitions of the word 'model' refer to the classical two- or three-dimensional scaled-down replicas of the toyshop or design office: only when it says 'something that accurately represents something else' does it embrace the mathematical models of which we are speaking. A model may be a set of equations, a computer program (or suite of programs), or something more visual such as a nomogram. The common characteristic of these models is that they enable the behaviour of real-life systems, such as manufacturing plants, transport networks, or even thinking or speaking animals, to be wholly or partly simulated.[7] To be useful, this simulation must normally be cheaper or quicker (or both) than the real thing, so that exploration and prediction of unknown and new situations can be performed more usefully with the model than by a real-life trial. To be useful, the model must not simulate too inaccurately, though it does not have to be highly accurate in every detail. Indeed, the principal skill of modelling lies in achieving simplicity (and therefore comprehensibility and economy of effort) by concentrating on those variables to which the predicted behaviour is most sensitive. We should ask not what a model is, but rather what it does: a model's most important property is utility.

This utility may take several forms. The first is that of helping in *control and management.* A set of equations describing the flow of traffic in a street network with signals at intersections can be used to analyse the effects of traffic entry at any rate at any series of points and, if it is good enough, can help in the design of one-way street systems, linkage arrangements for traffic signals, or bus routes and timings—the objectives being optimum flow and safety. Computer programs which consist of equations defining the output from a chemical plant in terms of the operating conditions and inputs can be used to secure the adjustment of controls, manually or automatically, for maximum output or profit.[8] Simple equations describing the demography of a workforce (e.g. the linkages between engagement, retirement, wastage, and age-distribution) can be used to explore the personnel policies that could be most successful in securing various defined assemblies of skill and competence at specified future times.[9] The results of all these applications, however, will be useful only as long as certain boundary assumptions are not transgressed. Thus, the traffic model will have to be changed if a new flyover is built, the plant model if an improved catalyst is introduced, and the manpower model if economic conditions cause many more staff to resign.

The second form of utility, overlapping with the first, is that of *teaching.* If it can be shown and explained to someone how (and how far) a model simulates reality, he can then explore various policies using the model and learn where danger or benefit particularly resides. Since most management and control systems do require human intervention, this pedagogic use is important to get the humans in the system to play their part as well as possible. A well-known example of this use is the flight simulator: trainee pilots sit in a mock-up cockpit and move controls which are linked directly to a computer, which in turn tilts the cockpit and projects films of a runway on the 'windows'. Many lives and planes have been saved in this fashion.

The third utility is in the generation of helpful *conversation*. Much discussion is impeded by ambiguities about the meaning of words and the precision with which descriptions are made and understood. In perhaps its simplest form a model to aid conversation consists of a few words, circles, and arrows on a blackboard or flip-chart, drawn by one participant to convey his perceptions to the others. A mathematical model demands the precise definition of its inputs, after which the processing of those inputs goes on in a precisely prescribed routine. Consequently, a group of people discussing a situation on the basis of trials with a model have fewer problems of ambiguity, and therefore spend less time in arguing, perhaps fruitlessly, on the basis of misunderstanding.

But there is a countervailing misuse, and a dangerous one at that.[10] A model may look very convincing, and may be difficult to argue with, particularly if one tries to attack the underlying logic, which is almost certainly internally consistent. Bruised by an unsuccessful assault at this point, the hearer may be discouraged from penetrating to the analysis of the basic assumptions, and may therefore be led to accept the output of conclusions too readily. Equally, of course, a very good model may be used with the completely honest and open disclosure of assumptions to command the broad allegiance of informed opinion.

3. Some typical uses of models by technologists

(a) *Models in the disciplined factory situation*

Like *le Bourgeois Gentilhomme,* who was astonished and delighted to be told that he had been talking in prose all his life, the old-fashioned factory manager retiring this year might be equally surprised to know that he has always used a mathematical model. Yet he has, if he has been operating a standard costing system.[11] A standard cost sheet is a model: it discloses the planned or forecast cost of every resource used per unit output of product, including capital charges if wanted. It can thus be used to evaluate the effect on market competitiveness of granting a wage claim, making technological improvements, permitting a 5 per cent increase in throughput, substituting a cheaper raw material whose use drops efficiency by 1 per cent, or of making other possible changes. The factory consists of an assembly of cost centres, each one of which operates a number of standard costing comparisons, so that the factory manager is in effect supervising a hierarchy of models. He had common services—engineering, safety staff, analysts, medical effort— to deploy, and his own time to apportion. This he will do on the basis of priorities in which the monthly comparison of real performance against the modelled standard plays a key part (alongside consideration of safety, the law, health, and sudden catastrophe). For the sake of clarity, his model is probably updated only once a year, during which time real-life unit resource costs may wander appreciably from these used in his standard cost sheets: this is an illustration of the acceptance of known and limited imprecision in a model, in the interests of simplicity and comprehensibility.

Although the factory manager can command, and can order restoration of unit resource consumption to last year's level with dismissal of the plant manager as the price of failure, he will more commonly persuade.[12] For this purpose, his cost model is invaluable. He can demonstrate to his staff how quickly they will recoup their own salaries by the attainment of management objectives, and how powerful are the arguments for setting niggling worry A aside so as to strengthen the attack on major loss problem B. Better still, he can persuade everyone to join in his 'ownership' of the model, so that the month-end conferences cease to be trials, with the boss as prosecuting counsel, the operating staff successively placed in the dock, and the cost accountant acting as a sort of police informer. A factory that has taken its standard costing system as its main source of wisdom will be a united factory that can practise management by exception and management by objectives to great effect.

However successful this may have been, no model user or modeller should ever remain content. Cost centre operations represent a balance of advantage, and although many improvement programmes genuinely increase factory margin, some do not. Sometimes such readjustments are helpful in drawing attention to uneconomic products that are being subsidized, and need to be dropped or corrected. Sometimes they are not helpful. If the factory processes are partly sequential, and also involve flexibility in the balance of products being made, then overall improvements may require sacrifice of profit at one cost centre to generate more at another. Since the standard instruction to a cost centre manager is 'maximize your profit', such a change is contrary to instructions and will be resisted. The situation is even worse if product demand and price fluctuate too rapidly for accurate response in the manufacturing system.

A case in point is a petrochemical complex, where crude oil is the input, and intermediates for making polythene, polypropylene, rubber, nylon, polyester fibre, together with motor spirit and fuel oil are the main outputs. Crude oil is a complex mixture whose precise composition varies from oilfield to oilfield. The main operations in a large refinery are first the unmixing of the components and secondly the conversion of some components into others by suitable chemical action. By adjusting the operating instructions, the proportion of all these products may be varied; greater latitude still may be obtained by capital investment in extra product storage. Not untypically, the annual reward for really sophisticated planning might be £0·25m—£0·5m for a plant complex with a capital value of £10m—20m. I.C.I. have published details of some of the procedures adopted for designing and operating a system for optimizing the operating margin and for setting priorities for further capital investment to improve the complex yet more.[13] The salient points are these.

First, it seems best to use a compatible hierarchy of models rather than one global model. The degree of detail needed is quite different for the best operation of one particular distillation column, the most profitable partitioning of a range of products between motor spirit blending and further chemical up-grading, and long-term capital investment planning. Although in principle this use of separate models could lead to sub-optimization, there are greater countervailing benefits for keeping every model as simple as possible. Thus, the model for ten-year planning is less detailed than that for one-year planning, which in turn is less detailed than that for day-to-day planning. What appears as a tactical output from the long-term model is written into the medium-term model as a strategic necessity, and in turn generates overriding operating rules for running the complex.

Secondly, the role of human participants must be included realistically in the model. No one should be given a boring or a superhuman job, and if possible everyone involved should have the interest of his job improved by any new management system introduced on the basis of the model. Thirdly, as in the case of a well-run standard costing system, everyone running the complex needs to feel partial ownership of the model. This can be achieved partly by teaching and even better by getting teams of the operators to work out parts of the model.

A particular benefit will arise if some of the model's numerical outputs can be translated into concepts. For example, an olefines cracker transforms a ton of naphtha into about 0·25 ton of ethylene (mainly for polythene), 0·15 ton of propylene, 0·10 ton of methane, 0·35 ton of motor spirit and various other by-products. It is the key starting-point for most major petrochemical processes, but is usually built to make ethylene, and is thought of as an ethylene plant, with valuable by-products. This concept leads quite naturally to the running of the plant in such a way as to optimize ethylene yield, i.e. to vary the naphtha feed rate up and down with ethylene demand. Some years ago, the running of a computer-based mathematical model using the technique known as linear programming showed this policy to be sub-optimal: more profit could be made by keeping the feed rate up and varying the furnace temperature so as to alter the ethylene output, i.e. by changing conditions to make more by-products when ethylene demand fell off. This policy, arrived at by some quite elaborate work, simply amounts to running the plant to produce motor spirit and treating ethylene as a by-product. This is a concept which is readily understood by everyone. Once this is done, the linear programming runs can be less frequent—but they cannot be abandoned, for the 'rule' may well be only an ephemeral one.

(b) *Investment models*

Technological change and market growth require new plants: how big should these be, and when and where should they be built? What processes should be used? Should one buy an expensive licence and build now, or develop a new process and build later, with loss of market share but with better prospects for earning money elsewhere with the new technology? Should one aim always to have spare capacity or should plants be run as nearly full as possible (with some risk of short-age)? All investment decisions, in retrospect, could have been better; some could have been so much better that the history hurts. Some, by creating chronic overcapacity, can upset the profit of an entire industry on a world scale, deprive it of the cash needed for further extensions on the due date, and set off a particularly violent and harmful swing of the trade cycle. How can performance be improved by taking thought before the event? And how can the improved procedure be kept simple enough to avoid the ever-present danger of 'paralysis by analysis'?

In essence, all investment models are methods of logically combining a whole range of diverse information about how much of a particular product might be sold, what price it might fetch, and how much it might cost to put up a plant and produce it.[13] Virtually all this infor-mation is subject to a degree of forecasting error which can on occasions amount to monstrous self-deception. Fashion changes have swung procedures between different models. All of these are useful, but the lazy or ignorant man's desire to use one economic merit index, and one only, deprives him of an opportunity to think.

The simplest decision model uses the financial rate of return (profitability) that will be obtained when the new plant is full. For the first year when this is so, estimates are made of the likely prices and the likely costs; the margin, suitably reduced for overhead contributions, is then divided either by the full or the depreciation capital value to give the return on capital, which may be computed before or after tax. This return can then be compared with alternative policies of building other plants, putting the money in the bank, and so forth.

A more elaborate and widely adopted decision model is that using discounted cash flow analysis.[14] This makes it possible to study the time taken to recover investments with different profiles of spending and reward. If products have short lives, calculations of this sort are necessary, and if the plant can only make one possibly short-lived product they are vital. This model can advantageously be displayed graphically for various different sizes and timing of plant, for this throws emphasis not only on the time taken to recover the investment, but also on the duration and depth of the maximum commitment and risk. If D.C.F. is used only to rank projects in terms of rate of return for a specified payback period, or different payback periods for a specified rate of return, much useful information is ignored.

D.C.F. analysis, however, is no more than the simplified limiting case of an investment model that takes into account a variety of features of a business. Such a model can be made more complicated almost without limit, and in its over-elaborate forms it can build in estimates of the probability profiles of any or all of the constituent numbers. Such probability profiles are usually conservative guesses and no more, and their cumulative effect (graced by the title of 'risk analysis') is usually to make the proposal less attractive and give a spurious appearance of precision to some doubtful economic indices that are much better recognized in their truly approximate colours. What is much more useful is the analysis of the sensitivity of the final economic index (rate of return, or whatever else) to various reasonable specific hopes or fears about levels of different variables: harsh competitive pricing policies, late start-up, better materials efficiency and so forth.[15]

If obsolescence is very slow indeed, because products are mature and innovation difficult, then the difference between the results of analysis in terms of profitability at full capacity and D.C.F. will become small and, in the ultimate, will vanish. Where this is so, the rule is 'use the simpler model: it is harder to deceive or suffer illusions with simple sentences written in words of one syllable'.

The intelligent but untutored questioner (to whom we do not listen often enough) would ask searching questions of all these investment decision models, such as: How do you predict the selling price? How do you know how quickly to depreciate the plant? How do you know how quickly your research colleagues will improve the process and reduce costs? How do you know that your allowances for maintenance and breakdown are realistic?

On the question of prices and process improvement, the Boston Consulting Group have introduced some thinking that is very beneficial to the powerful producer, and usually to the consumer too.[16] Their basic thesis is that cost reduction is predictable, and depends on the cumulative production to date. This builds in a rough-and-ready version of the law of diminishing returns: every doubling of cumulative output reduces costs by about 20 per cent, so that the experience gained in going from a grand total of 5000 tons to 10 000 tons, from 10 000 to 20 000, from 20 000 to 40 000 and so forth, each slice 20 per cent off the real cost at the start of that phase. A plot of log (cost) against log (cumulative production) is then linear, with points for successive year-ends for a given plant bunched closer together (Fig. 4.1). The reasons for these benefits from experience are neither specified nor analysed: one year it may be a better catalyst; another year better control; another year better maintenance or a new kind of reactor. Empirically, the rule holds for an astonishing variety of products and, where profit margins change little, for prices also. Fig. 4.2 illustrates how such a model can be of great utility without any very firm theoretical base.

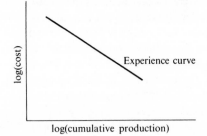

Fig. 4.1. Boston experience curve.

Costs predicted in this way for oneself and one's competitors can permit the formulation of pricing policy. If the product is fully patented with no competitors, and yet so useful or of such prestige that demand is insensitive to price, then the price is simply the figure that society will

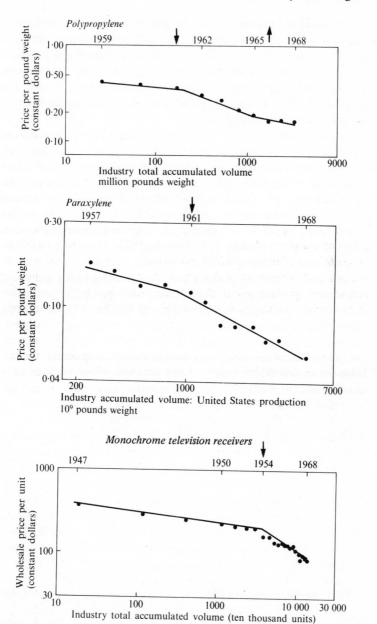

Fig. 4.2. Examples of Boston experience curves ↑↓, points of trade recession (↓) or recovery (↑).

(From *Perspectives on experience,* Boston Consulting Group Inc., 1970.)

not castigate as extortionate. If it is patented but price-elastic (i.e. demand would increase sharply if the price were lowered), there is a choice between high immediate profit on the one hand and long term strength arising from rapid progress down the cost-experience curve on the other, which will make it very difficult indeed for later competitors to catch up and overtake. If the situation is already competitive, it is possible roughly to predict everyone's margins from their positions on the experience curve. If the analyst is the farthest down the curve, he can either set a high price that will earn him more than adequate profits, and allow his higher-cost competitors to live with lower profits but a retained share of the market; or he can price so as to shave his competitors' profits to levels so unattractive that they give up, leaving him with a higher market share. It is possible to write this scenario for some years ahead, and model one's competitors' reactions to different .pricing policies, thereby exploring the price range that will permit one's own plant to run profitably at full capacity

The method may not work if there is severe overcapacity and a price war breaks out, although it may successfully deal with the situation after weak producers have ceased operations. In any event, it reduces the chance that over-optimistic pricing may lead to unexpected and unpleasant competitive reactions, and it should also discourage the building of gross over-capacity if the broad intentions of others are known. Developments of this model permit the study of the maturing situation, and of complete plant investment portfolios.

(c) *Human systems: conflict and manpower models*

We have already emphasized the vital importance of including people in models and in their use, and of avoiding the mechanistic merit and human indignity of the assembly line. It sounds undignified and inhuman to model human assembly and behaviour, and yet the reverse is the case. The perception of likely conflict should concentrate thought on means for its avoidance, and the perception of likely discomfort should suggest the need for different systems of behaviour. Finally, if conflict really does seem to be unavoidable, a conclusive demonstration that victory is inevitable may persuade one's opponent to climb down. The motto for all these models might well be, 'Forewarned is forearmed'.[17]

We have touched on the modelling of conflict in the Boston Consulting Group's approach to competition. Games theory offers mathematical methods for analysing a range of real-life situations, and often the very operation of converting a conflict into the appropriate symbols and numbers helps to define the essentials of the position.[18]

Such conflict modelling can be used to concentrate attention on the fundamental nature of economic situations that are giving trouble, and this, of course, brings us straight to the constructive art of the cartoonist. Consider three men bidden to share out £300 by majority vote. Any coalition whereby a particular pair agree to take £150 each can be upset by the excluded member offering a £160 share to whichever of the others will join him in an alternative coalition, after which the newly excluded member can seek to buy his way back by offering a £150—£150 deal to the recipient of £140. No arrangement is stable except a £100—£100—£100 solution, which is stable only because of the evident instability of all alternatives. This kind of model could, given a much higher level of general education, be used as a language in which to discuss the realities of pay claims and their relative merits in a situation of defined and limited increase in productivity. No other use of mathematical or alphebetical language can match the cartoon or the model in this case, and the funnier the cartoon or the simpler the model, the better the result.

We have so far seen models that can unite activities in a big factory and models that can clarify and perhaps avoid conflict. It can only be astonishing, therefore, that so little use has been made of them in the planning of education and employment. One reason may lie in the highly specialized character of education and the departmental structures both of educational and other institutions. Consequently, we are bedevilled by the narrow concepts that 'chemists do chemistry, and physicists do physics', which leads us to ask how much chemistry and physics there is to do in the departments so named within I.C.I., B.P., the D.o.I., and the Treasury. Presumably if the Treasury has no chemistry department, it does not employ chemists. In fact, as we saw in Chapter 3, chemists are increasingly employed outside their own subject, using its skills in other connections. Once such concepts of flexibility are introduced, manpower modelling takes on a new meaning. In particular, it becomes possible to use models to explore better long-term responses to changing circumstances. Instead of attempting precise analyses of predicted 'demand' on a year-to-year basis which shifts disconcertingly with the state of the trade cycle, one looks ahead for much longer periods, such as 10 to 20 years, and then writes scenarios covering a range of possibilities for long-term trends. Such extended time periods are made necessary by the mutual loyalty of scientists and technologists and their employers: a large company can expect to have kept half of each year's graduate intake after 10 to 12 years, and to retain one third of them until retirement.

One can then test the feasibility of different general policies for employment in particular firms or sectors of the economy, and examine whether there are enough opportunities for those educated as chemists, or as engineers, bearing in mind what we know of their expectations and versatility. If opportunities are insufficient for chemists as educated at present, we can examine options for educating more broadly, or for post-experience retraining. Conversely, if there are too many jobs for the available chemists, we can see how many of them could be done by historians.

The main manpower planning task for many technological units at present is to see how to move from steady staff expansion, which has occurred over the past quarter of a century, to stasis or contraction in numbers, at least in those sectors where maturity has arrived. Expansion is easy: mistakes can be ignored and left behind. Contraction implies an ageing workforce unless quite new patterns of fixed-period employment can be made acceptable. Unless internally consistent models can be devised and implemented, it is likely that every downturn of the trade cycle will be the occasion for redundancies, so that people are called on to find other jobs just when this is hardest. The natural consequences are that the able will not work in this part of the system, and that staff will not cooperate in the overall pursuit of efficiency and good operation. Modelling permits feasible solutions to be proposed, after which these must be discussed with and owned by everyone affected (see section 4 below). When this is done, operational models can be used to determine the actual flows, promotions, and recruitments that are needed.[9]

The decision to expand the provision of post-secondary education, until it embraces over 20 per cent of the 18-plus age-group, poses the opposite end of this problem.[19] If all places are taken up, it implies that one graduate will be surrounded, on average, by only about four (instead of about twenty) non-graduates. His task will therefore be that of skilled cooperation rather than automatic command. Graduate employment patterns must therefore be widened, incidentally providing a chance to get many jobs done better than hitherto.

(d) *Propaganda models*

We have seen that it is vitally necessary to concentrate discussion on the assumptions inherent in any model, and the more so if these are doubtful. It is important to recognize, however, that simplification is a game for any number of players. It is all too easy to lure society away from the essentials of a discussion, and spectacular outputs from a computer (and especially a large computer operating a large program suite) may command too much allegiance simply because they arise from a system of impeccable internal logic. The fact that the inputs are wrong, or questionable, may escape attention. Modellers must therefore realize that they may, however innocently, confuse and alarm the very people they are trying to enlighten.

Doom is a fashionable subject, and a graph emanating from the Massachusetts Institute of Technology that shows world population plunging from 6000 millions to 1000 millions in the year 2010 is good material for the front pages of the quality Sunday papers. All that is remembered afterwards is the result, and not that it was only one of a series of runs of an interesting model which, on other assumptions, can predict much calmer ways ahead with no holocausts. What is worse, insufficient attention is paid to the assumptions of the model, which include a very simple picture of natural resource exhaustion. This virtually ignores the fact that there are substitution strategies available for every raw material, which transform resource problems into energy problems and ultimately into questions about means for solar photon fixation. Moreover, the linkages between population increase and feedback into aggression are fascinating but arbitrary, and are open to challenge on many grounds.[20] One obvious point is that the equations are almost certainly different at different times, both in form and in detail.

To make these criticisms is the business of the recipient of a model output, and to take them seriously is the duty of the modeller. He has invented a new dialect of model language, and must both teach and listen. The Club of Rome has performed a signal service in publicizing the potential use of dynamic modelling, but society now has to settle down to learn the grammar and syntax of the language before it can properly use its statements. Although this sounds a mild enough observation, it embraces a great hazard. An unscrupulous politician will have no difficulty in persuading a set of honest university modellers to produce, in learned journals, models that show society torn apart by postulated consequences of immigration, taxation, dismantling of defences, wage claim restraint, or many other possibilities. Society has in the past engaged in mob reaction to much less plausible doctrines, such as those of the Nazis relating to racial purity or those of Marx concerning the ultimate withering away of the state. It is only a matter of time before demagogic politicians start using models. We must be ready for them with practice in the basis for legitimate criticism, or we shall be at the barricades very quickly indeed.

4. The relationship between modeller and user

We have explored in some depth, because we think it is important, a range of answers to the question 'What do models do?' We must of course also ask 'How do they do it?', but this book is in no way a substitute for the many professional and detailed texts which quite explicitly answer the latter question. We shall therefore concentrate on communication processes rather than on calculations as such.

The main utility of a model consists in the hearing of inputs too diverse and complex to be a satisfactory basis for action or decision about policy, followed by the speaking of clear and trustworthy output that is directly useful for these purposes. The processing interposed between the input and the output may involve the use of sets of equations, computer programs, or the scanning of graphs. It must be free from logical defects, and if possible should have built-in checking procedures. But the elegance of the mathematics, logic circuitry, or hand calculation must not deflect attention from the prime requirement of utility. There is some danger at present that professional concern about calculation efficiency, or about the breadth of the data input, may lead the model and its construction to become an end in itself. Modellers should not be too satisfied with the design of the means and too detached from the helpfulness of the ends. The job is a question, starting, 'What, if . . . ?'; the output is an answer, saying, 'if A, then B', B being a more useful message than A.

A further requirement is that the output B should not be unjustifiably cocksure. Having seen some disasters based on sophisticated mathematics done with unsound economic assumptions (ranging from national economic plans that could not fulfilled to investment decisions that led to profits much lower than those promised), many users are rightly suspicious of pat and precise answers. Where, as is usually the case, the inputs are imprecise numbers, or data involving some arbitrary assumptions, it is vital to be scrupulously honest about the precision or imprecision of the answers. This is especially important if the model is rather too complex to be understood by the average user, who consequently may fear being supplanted by a black box—even though his part in translating the model output into action may be quite vital.

One good technique is to use the model effectively to draw a map, i.e. to exclude solutions that are internally inconsistent or wildly improbable when a reasonable view is taken of the precision of the input. The remaining possibilities can then be presented as a series of options between which the user must decide. Thus, the running of a manpower model might result in the statement: 'With your declared view of your future staff needs, you cannot reasonably pursue your present personnel policy of engagement for unspecified duration; for if wastage remains at its present low level, you will be unable to recruit, will repeatedly be declaring redundancies during trade depressions, and will find yourself with an ageing staff of decreasing ability, for your good people will leave. Instead, you could meet your needs by devising contract terms that would permit you to recruit 50 per cent of your staff on 5-year contracts, or 40 per cent on 7-year contracts, or 30 per cent on 10-year contracts, or a suitable combination that the model will work out for you.' This then directs the mind of the user to his important task, namely to think of the terms on which he could recruit good staff on short service arrangements, and of the management problems of running a staff partly on long service and partly on short service. The presentation of probabilistic distributions of 'required service periods' would be more bewildering and less enlightening. Other ways are available of being honest about the uncertainties, but it is always possible to improve the comprehensibility (and hence the usefulness) of the answers. Unremitting attention to this is one of the modeller's key duties.

The complete system with which a modeller is concerned is shown in Fig. 4.3. The formal and logical links which to him seen over-riding,

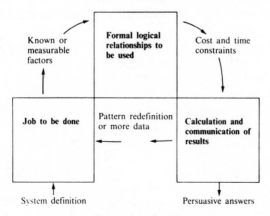

Fig. 4.3. The modeller's trilogy.

in that their development consumes most of his time, are simply one part of a trilogy; certainly no more important than either the problem definition of the communication of results. For the modeller himself is very rarely granted the executive power to act on the results of his model. He has a customer or customers (however broadly defined) who must in some way help him define the system under investigation, and who must share the assumptions at least in a qualitative sense if not in numerical detail. Only full involvement in this way will persuade the customer to believe counter-intuitive or non-obvious results if they later arise.

The arrows in Fig. 4.3 will be seen to form a closed loop, to which extra input and output arrows have necessarily been added. Most models to be useful have to be refined by successive iterations around the threefold way. Results from incomplete or oversimplified models are often the best method of stimulating discussion on where to go next. It might be thought that pure scientists would therefore make poor modellers, accustomed as they are to searching for factual frameworks and then reporting their work in the polished, historical, almost unassailable format of the scientific paper. Fortunately, the scientific paper (which is so necessary to the progress of science) is not a true representation of how scientists actually think and behave; in practice scientists and technologists make very good modellers so long as they give due importance to communications.

The idea of a compatible hierarchy of models, instanced in the petrochemicals complex earlier in this chapter, is an important and useful one. It is essential to avoid confusing one customer by using an overlarge model when he is interested only in a small but detailed part of the output, or another by using a minutely accurate model to display broad strategic outlines. Both will certainly be costly and probably ineffective. Artistic simplicity, though difficult to describe, is the ideal that should be sought. In terms of Fig. 4.3, some of the 'known or measurable factors' on the top left-hand side could well come from another coarser model; or some of the 'persuasive answers' of the bottom right could be fed into a still more detailed model, remembering in both cases that the inputs labelled 'system definition' must be internally consistent. The process can be compared with that of an automobile manufacturer, who would certainly go out of business if he tried to sell the same car to everybody; instead he uses common design ideas, and even components, over several cars in his range in order to satisfy the diverse needs of his customers.

5. Some hazards in modelling

The models mentioned below are not chosen to illustrate methodology (e.g. simulation, linear programming, manpower, capital allocation, stock-holding). The inventive modeller in the right place at the right time will see these opportunities, and seize them if he can. In any case methodology and application taken separately describe only single parts of the trilogy we are concerned with. The following examples are illustrative of possible pitfalls in the areas of overlap of two or even three of the components.

(i) *A 'wrong' model.* If one of the equations in the model is of a radically different form from the true behaviour in the system being modelled, then the model has neither explanatory nor predictive powers. If Newton had proposed that gravitational attraction depended on the inverse cube of the distance, then there would be a demonstrable mismatch between his model and the known solar system.

(ii) *An incomplete model.* This is more insidious because the modeller has apparently built in all the logical relationships needed to answer the questions posed about his system, and therefore his answers must be 'right'. Thus a naive model of traffic flow through a tunnel, based on non-turbulent liquid flow through a tube, might predict that higher traffic speeds would always mean greater throughput of vehicles. But a more complete model would take into account the increased distance between vehicles at these speeds. This has been measured for a New York tunnel, in which the speed for maximum flow is 19 mile/h—no greater, no less. But it would be pointless having completed the model to apply it to an incomplete system: if the true rate-limiting step is toll-collection, strict imposition of a 19 mile/h speed limit benefits nobody and is the 'wrong' answer.

(iii) *A misdirected model.* Many processes which involve the interactions of diverse elements through time are stochastic, i.e. subject to chance events. Because the system in real life has randomness built in, there is a temptation always to make the model equations and the communications medium reflect this. A long-term marketing model for a new industrial chemical which takes account of product substitution effects, the level and timing of probable toxicity legislation, possible competitive action and so on, is merely confusing to the marketing manager if it presents too wide a range of answers. Initially, at least, the simple but logically correct combination of the most likely values of the variables will tell him his approximate sales, which is all he wants. Conversely, a study of stone transport in a quarry, from mechanical shovels at the quarry face on large trailers to a central crushing unit, must explicitly consider random events, because it is such events which determine the growth and decay of queues at the shovels and the crusher. Stochastic versions of both the market and the quarry model are 'true', but in the former case a deterministic model is more relevant (and cheaper).

(iv) *A spuriously precise model.* All numbers coming out of models should be treated as approximate until they are shown to be accurate. If a piece of machinery breaks down once a week, not only is it unreliable but there is also a danger that the daily chance of a breakdown will be printed out by the computer as 14·2856 per cent. All three areas of interest are involved here: the system provides base data that vary in accuracy and precision, a computer will calculate precise answers that are accurate only if the equations are true, and the communication of the answers can be as imprecise as the modeller chooses.

(v) *An inaccurate model.* This type is dangerous because it can remain undetected. A classical example is in manpower models for young professional staff, where the inaccuracy if the model is based on the wrong variable. The real system shows that the chances of an employee resigning voluntarily appear to depend strongly on age. An obvious model for predicting future staff losses is thus one in which measured past age-dependent wastage rates are applied to the present age structure. This works well until a couple of years of high or low recruitment, or a change in the average age of recruits, reveals the fallacy. Wastage behaviour is in fact better predicted by length of service, and it is only the close correlation between age and length of service which confuses the issue. So the more robust model is one in which the historical wastage—length of service curve is applied to the present length of service distribution, though even then we are making no allowance for behavioural changes which might be induced by external economic factors.

6. Models as a new kind of language

Properly used, models can provide the technologist with a new means of communication. To some extent, they combine the strong features of alphabetic and mathematical language (see Table 4.1). Alphabetic language connects words from an enormous dictionary by logical operations that are simple. As soon as descriptive language is applied to

Table 4.1. Languages

	Vocabulary	Syntax	Perception
Alphabetic	Wide	Limited	Immediate
Mathematics	Narrow	Extensive	Delayed
Models	Wide	Extensive	Useful and dangerously immediate

logical connections of any appreciable complexity, the reader gets lost. This is partly because such language is clumsy for this purpose, and partly because its predominant use for simple logic leads to the expectation of instant comprehension and rapid reading, so that sophisticated logic gets skipped. Mathematical language conventionally performs very sophisticated logic on words from a rather impoverished dictionary. But it carries with it the expectation of a need for effort in order to comprehend. Consequently, being slow and toilsome to use, it does not become general currency.

Models can employ a large vocabulary of words and can incorporate quite complex logical operations which, however, are summarized and assembled in a manner that encourages and permits rapid scrutiny, with quite good comprehension of the general drift. Sometimes this superficial comprehension is too convincing, and leads to an illustion that the basis for the output has been well scrutinized by the hearer. But, provided that this temptation can be avoided, the man in a hurry using models can work easily with much more sophisticated logic than he can with descriptive language alone. Models therefore have the power of a new language, and in just the same way they open up the possibility of embarrassing or costly misuse.

Models, then, are used to understand, control, and improve a wide range of activities which bring together people and things that interact. To make sense, they must be used; otherwise, however elegant, they do not undergo the evolutionary development that results from natural selection. To begin with, modellers concentrated on the study of the things: now increasingly they recognize that the proper study of mankind is man. At this point, there is a conflict between the temptation to manipulate and the duty to converse, communicate, and persuade. The first, unacceptable, route leads into Aldous Huxley's *Brave New World* or Orwell's *Nineteen-eighty-four* (which is now quite near); it can embody steps such as the actual use of librium on a large scale for maintaining prison discipline, or proposals for issue of valium to teachers for maintaining classroom discipline. The second acceptable route requires the devotion of great effort to patient teaching and learning: the rewards are high but the process is difficult and often tedious. We believe that technologists must not be deterred.

5

Constraints in the use of natural resources and energy

1. Introduction

This is the longest and most complex chapter in the book. It is inescapable, however, that this whole area of constraint is a single system, albeit with many facets. We have therefore been compelled to treat it in a single chapter. Our task would have been far more difficult if we had been restricted to the use of descriptive text alone; hence we have relied heavily on a hierarchy of cartoons and interlinked models, doing our best to take the advice we ourselves preferred in Chapter 4.

Over-simplification has been inevitable. Our aim in this chapter is to help the technologist to distinguish real research problems and opportunities from mere scare-mongering, to recognize constructive social attitudes, and to respond helpfully. Even a superficial treatment should enable the humane technologist to pick out some important key points that indicate where concentrated effort is most needed. If in doing this we have occasionally oversimplified complex situations, we beg forgiveness: nevertheless we believe that our main framework, and the broad conclusions to be derived from it, are accurate and important.

The total system

Fig. 5.1 is a schematic diagram of some of the more important existing material and energy flows in our technological society. For the sake of simplicity, many links are not shown.[1] In particular, there is no indication of the capital investment and ongoing energy requirements that are needed by virtually all the processes in the diagram. It is important to remember that this diagram is no more than a cartoon, one viewpoint among many. An alternative way of representing some rather similar concepts was given in Fig. 2.1 in Chapter 2. Yet a third picture could be devised starting from the well-known carbon cycle, for instance.

Before trying to understand just where and why constraints appear in this system, we propose to describe it as it exists now in rather more

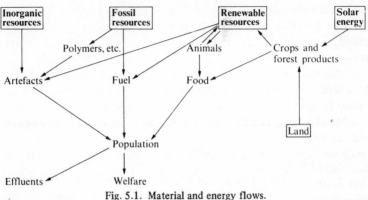

Fig. 5.1. Material and energy flows.

detail. Inevitably this involves us in history, i.e. in what the diagram used to look like two hundred or more years ago; and from that we make some projections as to what is might look like in perhaps a hundred or so years' time. Only the links change—the main elements remain.

2. The state of the world

(a) *Inorganic, fossil, and renewable resources*

It is convenient to consider natural resources, including energy, under three main headings: inorganic, fossil, and renewable. Inorganic resources include such things as limestone, iron-ore, and all minerals and aggregates, not forgetting liquid and gaseous inputs like water and air. Typically they are non-renewable, although recyclable. The total reserves of inorganic materials are enormous, although the easily accessible, highly concentrated minerals may arise from infrequent geological accidents and may be quite limited (Table 5.1). A high proportion of

Table 5.1. Reserves of metals in the earth's crust (to depth of 1 km)

	Weight (tonnes)
Silicon	$3\cdot8 \times 10^{17}$
Aluminium	$1\cdot1 \times 10^{17}$
Iron	$7\cdot1 \times 10^{16}$
Titanium	$1\cdot2 \times 10^{16}$
Zinc	$1\cdot0 \times 10^{14}$
Copper	$8\cdot0 \times 10^{13}$
Silver	$1\cdot4 \times 10^{11}$

Source: F. Roberts, U.K.A.E.A. (unpublished)

the world's aluminium and phosphorus—to name two wanted materials— is to be found in inconveniently low concentration in rocks of high chemical stability so that extraction requires considerable energy, and is costly.

Fossil reserves are mainly used for fuel, but more recently the route to artefacts via polymeric materials has grown significantly in importance. In effect mankind is tapping long-gone solar energy. The total reserves of fossil carbon sources are highly concentrated and convenient, and are largely confined to relatively young rocks. They are of course being used rapidly. Within a few decades, the most conveniently pumpable fossil oil will become increasingly difficult to find and extract, so that costs will rise to the point where the extraction of hydrocarbons from shale and tar sands, and the conversion of coal to hydrocarbons, become economic. Thereafter (Table 5.2) costs will continue to escalate as progressively poorer sources have to be worked.

Table 5.2. Hydrocarbon extraction costs

Source	Production costs ($ per barrel of oil equivalents)
Persian Gulf: primary sources	0·15–0·25
U.S. secondary sources of oil exploited using detergents, etc.	3·5
U.S. shale oil	7·05
U.S. coal-oil-gas refinery	5·70
U.S. coal-derived methane	6·20

Renewable resources consist of recent vegetable and animal matter, generated ultimately by recent solar energy (although still using a very small proportion of that which reaches the earth's surface). There will probably be an interesting balance, perhaps fifty to a hundred years from now, between the economics of hydrocarbon recovery from fossils, on the one hand, and conservative hydrocarbon production from renewable vegetable sources, or more direct use of carbon dioxide and solar energy on the other.[2] Renewable resources, it is important to recall, were almost the only source of fuel until about three hundred years ago. With historical perspective, the fossil carbon-using age will be seen as a brief interregnum of about five hundred years—equal to the life-span of the Roman Empire. During this period man will have learnt to utilize inorganic resources and to develop technology so as to harness photosynthesis, photosynthetic chemical and material production, and possibly fusion as an inorganic terrestrial energy source, at a productivity several orders of magnitude higher than that of classical agriculture. Man will also, as in the diagram, have used almost all of this extra capability to increase human population and thus set very severe problems of congestion, organization, and control of aggressive behaviour.

With this historical perspective, the development of chemistry based on hydrocarbons or on carbon monoxide plus hydrogen can be seen to be on a steady course, being steered by scientists and technologists who are in close rapport with the biologists, materials technologists, doctors, agriculturists, and engineers whom they serve. The technology required in the short term for extracting the required hydrocarbon or synthesis gas from coal and oil is receiving a great deal of attention. But there is a major need for a completely new look at the renewable-resource technologies, for despite great increases in productivity our present agriculture and forestry industry cannot be right.

First, it is sited where climatic conditions happened to encourage primitive man to be vigorous, organized and educated. It is now possible, by suitable technological adaptation, for vigorous, organized, and educated activity to be deployed in many more places. Accordingly, high-productivity agriculture can move from temperate zones, where solar energy supply varies from poor to moderate, nearer to the equator, where solar energy supplies are good. Israeli agriculture provides an elegant demonstration of the possibilities.[3]

Secondly, the crops grown at present have nearly all been chosen arbitrarily and then bred on quite a narrow genetic base, again for temperate zones. Ecological control problems for new crops in new areas will be severe, but they will be soluble by conservative means (which may or may not include the present type of chemical 'pest control'—itself possibly a model wrong and misleading because of its concept of antagonism). The processes of selecting new crops, learning about them and breeding better strains will take time and ingenuity.[4]

Thirdly, agriculture has not been designed to use the whole plant (or tree) to maximum advantage. Strategies are needed for maximum recovery of protein, usable cellulose or other carbohydrate and oil, coupled with microbial or enzymatic upgrading of that part of the plant that is not directly usable. Such upgrading may be to convert roots, stems, cellulose, or starch to oil, protein, or simple chemicals like ethanol that can be used for fuel or chemical synthesis.[5]

Finally, agriculture has not been designed to recognize that fossil hydrocarbons will increase in cost and decrease in availability, or that capital will be scarce and expensive. These considerations will greatly affect the choice of fertilizers and rates of application, and strategies for tilling. Essentially we shall be working towards a supply situation, as in Fig. 5.2, in which fossil resources will be of little or no importance but where increased use is made of solar energy as a result of these changes.

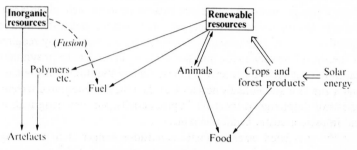

Fig. 5.2. Renewable resources.

It is fortunate that several decades are available for a start to be made on this second agricultural revolution, which will probably be the most ambitious co-operative enterprise man has ever attempted. It will pay off as it proceeds, and early successes may, as usual, produce new conservatisms that will delay further progress. Because we are probably speaking of a process that will continue for five hundred years (given a favourable political environment), it is impossible at this stage to do more than indicate some initial options for economically and politically acceptable first steps, together with ideas on general direction.

(b) *The onset of constraints*

The recent and rather sudden transition from world complacency about natural resources to world concern (and perhaps neurotic overconcern) is, of course, the result of political events rather than any sudden technological or geological revelations. Essentially, the secondary manufacturing countries have, over the past hundred years or so, used their military and economic command over trade, technology, and transport to preserve advantageous trading relationships with primary producers— whether of food, materials, or energy. The arrival of independent and effective statehood in the primary producing countries, coupled with the education of their citizens and the development of international bodies where common interest groups could coalesce, often as a reaction to the developed bloc, has made an increasing number of able Africans, Asians, Canadians, Australians, and New Zealanders aware that they had trading strength that was not being used. However, it had been much easier for the manufacturing countries to unify their strength through the operation of commodity exchanges than for the producers of meat or minerals to recognize and then capitalize on their strength to the extent of trading communally, setting aside beggar-my-neighbour devices. Sometimes there were emotional ties (as in French North Africa and the British Commonwealth) to encourage separate and divided trading by the primary producers. Genuinely concerned and able colonial civil servants and traders had seen themselves as the architects of a mutually beneficial partnership with the primary producers. They failed to see the extent to which this partnership would in due course operate against the colonial economies; cheap food was the counterpart of cheap manufactures, in this best of all possible worlds. Whether their efforts would have lasted longer if they had not been accompanied by less enlightened traders, such as those who made a family fortune by two round trips in the triangular trade of salt and cotton goods, slaves, and sugar and raw cotton between Britain, West Africa, and the Caribbean, it is difficult to say.[6]

It is odd that the unnatural trading balance was undone almost simultaneously by various events, including the warlike gathering of the Arab oil producers because of the astonishing military and economic performance of the small state of Israel, and the more turgid defensive response of the British Commonwealth countries to U.K. entry into the E.E.C. Oil had for many years been seen to be a commodity of growing scarcity value; protein had not. A variety of factors, such as drought in Africa, poor harvests in Russia, and the rapid economic growth of Japan produced a protein shortage in 1972–3 which coincided with the arrival of unprecedented agreement among the oil-producing and exporting countries (OPEC). Until that time they had never held their unity together for long enough to grasp the economic power that was theirs for the asking, for 60 per cent of world oil reserves were in the Middle East, and the United States was becoming increasingly dependent on imports. Consequently the price of oil and protein trebled, almost as though they were indissolubly linked, and the much more disparate and dispersed producers of phosphates recognized and took a like opportunity.

Thus, the initiative was simultaneously seized by the owners and controllers of the supply of inorganic, fossil, and renewable resources alike: even British coal miners joined in. Their initial display of strength upset economic equilibria, which, as always, had depended to some extent on faith, confidence, and illusion. Money became scarce because no one knew what could safely be done with savings. The initial loss of confidence by the investors in manufacture naturally spread to the investors in new supplies of raw materials, the winning of which is even more capital-intensive. What now has to come is a renegotiation, in terms of a new common interest, between equally powerful resource owners and users; yet this requires political machinery that does not exist.

(c) *'Exhaustion', a wrong and misleading model*

It is frequently stated, in newspapers and elsewhere, that oil (or some other resource) will be exhausted by such-and-such a date. (The year 2000, being a round number and easily remembered, is often found suitable.) As a warning that we are using oil heedlessly and mindlessly, and should pause for thought about the use pattern, such a statement can fulfil a laudable purpose. As a model that can guide actual policy, it is dangerously misleading and inaccurate. For, we repeat, the onset of constraint is political, not geological.[7]

There will never be a moment when a group of men buy themselves pints of beer and say solemnly, 'Well, we pumped the last ton of crude oil today (or mined the last ton of copper). It was good while it lasted, but now, what next? Bottoms up, chaps.' Although this caricature appears ridiculous, exactly this sequence of events *does* arise in some specific situations: for example, that of an automobile plant, so that the last motor car ever to be made in that plant can be seen, and touched, and nostalgically driven. People tend to move modelling concepts from situations in which they are valid to those in which they are invalid, so that this elementary word of caution is not superfluous.

The counterpart of the 'exhaustion by year 2000' statement is the 'world requirement by year 2000' statement derived from the linear extrapolation of a semilogarithmic plot of world consumption over the period 1900–70. The semilog extrapolation, presupposing exponential growth, carries the important proviso 'assuming that preferences and relative prices do not change discontinuously'. Over the period 1900–70, preferences and price differentials have both changed no more than gradually, so that there has been a general exponential pattern of growth (although different outputs have grown at different percentage rates per year; e.g. 0–2 per cent for populations, 10–25 per cent for plastics and synthetic fibres). Now, however, we are almost certainly facing large and relatively more sudden changes in price and preference patterns. Since a statement about 'exhaustion by such and such a date' must, if it is honest, be based on an allocation of a remaining reserve stock among year-by-year demands, projections of future use of requirements must be a necessary component.

Reality is more complex and secure in several separate ways. All resources occur in a variety of conditions, from which extraction, transport, and processing vary in difficulty and cost (both in energy and capital). Current supply patterns reflect a part-conscious optimization of required material in the ore or raw liquid in which it is contained (e.g. combustible hydrocarbons in pitch); amount of spoil to be moved per ton of ore; value or disposal cost of spoil or contaminant (for pitch can be sold for roadmaking); and cost of movement of product to point of use.[8] Even if we have an encyclopaedic and total knowledge of the accessible part of the earth's crust (say, down to a depth of five miles) the calculation of the annual rate of offtake would be a formidable undertaking, requiring knowledge of energy, labour, and transport costs, as well as current interest rates on capital. But we would also have to remember that prices would be expected to rise as scarcity became certain. Nearly all resources are price-elastic in some way or other; demand may not fall steadily as the price rises, but it will begin to fall substantially as a substitution option opens up. Consequently, demand would fall off as reserves diminished.

Further, we do not have anything approaching this knowledge of the earth's crust, and the rate at which we increase our knowledge (which at present is almost pathetically minute), depends mainly on economic factors, and in particular on world interest rates. Such information is costly to acquire and takes a long time to utilize, even if it reveals mining possibilities that would immediately be economic. Consequently, if society takes an overall decision to consume, rather than postpone consumption so as to save and invest, exploration is one of the forms of technological activity that inevitably suffers. This is true even if resources are nationalized. While some exploration may be undertaken on general arguments about national or international interest, governments have only limited supplies of capital, which they must allocate among medical, educational, and social services; exploration must take its place among these. Table 5.3 shows the nominal 'dates of exhaustion' of proved reserves of various resource products, calculated in 1971. Not surprisingly, they are close together and about twenty years ahead, for exploration farther ahead than this can be represented as a waste of capital.

Table 5.3. Dates of exhaustion of known reserves

	Date of exhaustion
Uranium-235	1985
Tungsten	2005
Copper	2000
Lead	1987
Zinc	1988
Tin	1990
Gold	1988
Silver	1988
Platinum	1988

Source: P. Cloud, in *Environment*, ed. W. W. Murdock (1971).

(d) *Energy and food*

We have seen above that the system will not let us go on as we have been doing, but that simple exhaustion and collapse is not a likely prognosis. There remains a vast range of intermediate outcomes, involving adaptation and innovation as mankind learns to change some of the links in the system. Two of the most important areas to be tackled are energy and food.

Until about fifty years ago, there was a simple relationship between the nature and the uses of resources: most materials for construction, engineering, and domestic objects came from inorganic sources; most energy came from renewable resources before 1700 and has come from fossil resources since; and all food, clothing, and packaging material came from renewable resources (Table 5.4).

We are now in a position where we can meet any requirement from any of the three sources: we could even make food from wholly inorganic sources by growing *Chlorella* using carbon dioxide from carbonate rocks and using light generated from nuclear power! However, as we have seen in Fig. 5.2, in another one or two centuries we may again be producing all food, clothing, and packaging from renewable resources, although fibres, fabrics, and films could still be made from simple organic building blocks such as ethylene, derived, not from crude oil, but from ethanol made by fermentation. Timber and cellulose products will probably increase in importance, and the supplies of inorganic materials will be constrained by the costs and availability

69

of fuel and capital. Partly for this reason, and partly because of increasing technological design for recycling, it seems highly unlikely that shortages of inorganic ore supplies will be serious.

Energy. Energy is best generated extra-terrestrially; indeed, if the earth were not almost precisely where it is relative to the sun, the earth's surface would be too hot or too cold for life. It is, however, convenient to have tactical terrestrial heat sources to enlarge the habitable areas of the globe. Moreover, it is not easy (and requires a great deal of capital) to convert solar photons directly into the work necessary for a technological society. At present, hydroelectric equipment is the only large-scale means available for doing this. Tidal power would make use of a further extra-terrestrial energy source, gravitational attraction, but the potential output is not high and capital costs are large. As the costs of terrestrially generated energy rise we may expect the following socio-technological trends:

(i) There will be straightforward economizing measures, such as better architecture for using solar energy and better insulation. The development of the second agricultural revolution will lead to more cities and habitations being built nearer to the equator; when this occurs, architecture will be concerned with the development of non-resource-intensive cooling.

(ii) The use for heating of electricity, generated using heat engines (with the waste of over 60 per cent of the fuel value as low-grade heat), will diminish. More direct methods of heating will be used: these could still be included the use of electricity derived from fuel cells, which have much higher thermodynamic efficiency, but at present are too capital-intensive.

(iii) Electricity demand for power uses will continue to increase, but the rate of increase will probably be held back to some extent by technological developments in energy economy that are motivated by high power costs. As an example, shaping of fabricated objects by moulding, extrusion, and precision casting will become increasingly attractive, so that turning and milling will diminish in importance. But electricity is likely to remain the main source of power and light.

(iv) The growth in the consumption of gasoline and kerosene for automobiles and aircraft will continue, but at a rate diminished in due course by cost increase, and more significantly by urban congestion and by increasing use of large-scale vehicles for public transport. The real hope is that education will enable telecommunications to replace

Table 5.4. Nature and uses of resources

Source	Construction, etc. material	Energy	Food	Clothing, packaging
Inorganic	All (metal, glass, brick ceramic) → 1920	Nuclear, 1955 →	None	Metal, 1900 →
Fossil	None → 1930 Plastics, rubber, 1930 →	Very little → 1700 All, 1800 → 1955 Most, 1955 → 2000	None → 1980 Protein 1980 →	None 1940 Synthetic fibre and film, 1940 →
Renewable	Timber, always Rubber, 1900 →	All → 1700 Very little 1800 → 2000	All → 1980 Most 1980–2050 All or most 2050 →	All → 1920 All (some via chemical synthesis) 2050 →

Dates indicate start or finish of trends shown by arrows.

some of the more fruitless personal movement undertaken in pursuit of affairs.

(v) Alongside all of these moderated rates of increase in Western society, there will be a steady increase in energy usage in undeveloped countries: as Table 5.5 shows, a levelling-up in energy usage in all countries to that existing in the U.S.A. in 1970 would increase the world total by a factor of five. With increasing populations this factor will be even greater. It seems likely that this state of affairs will not be reached for a very long time, if only because of the capital requirements; these would be far greater than those for the U.S.A., where much of the capacity is in the form of cheap equipment for burning oil or gas, built when steel and other energy-dependent items were manufactured with cheap fuel.

Table 5.5. World energy consumption

	Total consumption (1970) (10^{12} Btu)	Consumption per capita (10^6 Btu)	$\dfrac{\text{Energy } per\ capita}{\text{U.S. energy } per\ capita (\%)} \times 100$
U.S.A.	67 444	329	100
Canada	7039	329	100
Western Europe	47 874	134	40
U.S.S.R.	31 994	132	40
Japan	11 255	109	33
Africa	3677	10	3
Asia	9564	8·5	2·6
World	214 496	59·4	18
World consumption adjusted to U.S. per capita level	1 188 312	329	100

Source: *Energy in the 1980s.* Royal Society (1974).

It is difficult to say what role can be assigned to nuclear energy, with its extremely high capital and safety requirements. Fission power offers a further respite of up to a few centuries (depending on the extent of population control and energy economy), but only at the expense of a large investment, not only of capital but also of energy itself. Calculations are at present being undertaken of the net energy flow for various scales of building programme, but it is difficult to visualize the nuclear power industry otherwise than as a consumer of energy from now until the end of the present century. We shall return to the options later in this chapter. If it turns out that the first century of fusion power development is as energy-hungry as the first two decades of fission power development, there might seem to be a commanding case for concentration on solar energy and agriculture, with a strict rationing of the energy input to nuclear development. This would also entail careful consideration of the availability of fissile materials at various levels of ore concentration, for it is imperative that the fission power interregnum should be maintained by fuel, extraction technology that operates at a handsome overall energy profit, after allowing for isotope concentration, spent fuel treatment, and breeding.[9]

Food. There is no shortage of land suitable for growing food or produce for fuels or materials, provided that world population is made to level off in the region of ten to twenty thousand millions. Further, there is no lack of solar energy; botanical conversion efficiency lies in the region of 1—5 per cent. Nor, for the present, is there a serious energy limitation on advanced agriculture. In the U.S.A., which produces a very large agricultural surplus, only 3 per cent of the national energy consumption goes into agriculture, and this includes the energy and hydrocarbons consumed in fertilizer manufacture. Indeed, transport (16 per cent of the total) and domestic heating (28 per cent) constitute far more important areas for economy. While it is true that intensive agriculture may consume more than one fuel calorie to produce a food calorie, we can see from these figures that man uses his food calories in a much more discriminating, purposeful and elegant a way than he uses his petrol or gas: he uses about 100 fuel calories, for all purposes, per food calorie.

Food shortages arise from civil disorder, poor social and educational systems, and climatic fluctuations (which are much more serious in primitive communities, where agricultural methods are more sensitive to bad weather or drought). Rates of well over 3 per cent annual expansion in agricultural production—a level that would produce an expansion in *per capita* production everywhere—are attainable by present methods, given steady increase in fertilizer application (requiring investment), husbandry (requiring education), and irrigation (again requiring investment). Sophistication is also required in ensuring a correct balance in minor components—essential amino acids, vitamins, and trace-elements—especially where diet is near the margin. As usual, there is less need for education in communities that are rich and therefore educated, and more need for investment in communities that are poor and therefore cannot save. Since existing agricultural practices, as already adopted in the most advanced countries, could feed about sixty thousand million people if applied to all the suitable land, it is congestion and consequent aggressive behaviour rather than starvation that basically determines a maximum acceptable population.

In this situation, agricultural technology has in fact been held back less than urban industrial technology by human, environmental, or resource factors; the chief difficulty has been low prices where productivity could rise most (like the American Middle West). The main factors stimulating agricultural development have been government price-support policies (usually started in wartime) that have provided investment for equipment, amd migration of labour from the land that has protected agriculture from problems arising from wage inflation, by enforced increase in manpower productivity.

Economic imbalance, and rigidity of dietary traditions in poor countries, have played a further part in separating the world into areas of surplus and shortage. When technological advance can be utilized, there are many lines for attack.

(1) Higher-yielding crops can be grown in suitable areas, consideration being given to the use of the whole plant (of which about two-thirds is at present discarded) to yield starch, cellulose, protein, and oil.

(2) Fermentation techniques can be used to convert excess cellulose into food.

(3) Animal protein consumption can be reduced—if need be, to zero.

(4) Animal protein production can be made more economical in terms of feed inputs, by factory farming (although in the long run this approach is probably unacceptable).

(5) As a temporary measure, food can be made by fermentation of products (methanol and paraffins) from oil and gas.

(6) Plant breeding and genetic transplants may greatly increase the range of plants that fix nitrogen.[10]

All of these measures would cause changes in husbandry that would call for new techniques to control pests and plant disease.

(e) *Present preferences*

The design of future plans and policies depends not only on an awareness of the general state of the world at any one time, and of the nature of the economic effects of resource utilization, but also on a knowledge of the habits people are developing. These habits may have been mindlessly acquired, and man may be in need of abandoning some or all of them, but it will almost certainly be easier to reach conservative and conservationist behaviour in some fields than in others. Programmes of reform are often more successful if taken stepwise, so as to avoid creating large-scale social antagonism. Confidence in the possibility of the easier reforms, born of actual success, can help people to tackle the more difficult ones. What, then, is the nature of our consuming habits, when and where is it best to begin changing them, and how can this be done?

Evidence is available from the level of consumption *per capita* of various items in various countries, and also from the rates of changes in these figures over the past decade or two. Tables 5.6–5.10 show the U.K. situation, which presents a remarkably clear picture.[11] There were five ranges of growth rate during the period 1961–71.

(i) Products that increased by 110–170 per cent during the period, namely, plastics, man-made fibres, petroleum, salt, and fluorspar. These are the items representing the saving of labour; in retailing and domestic cleaning and maintenance, reduction in the care of clothes, reduction in carrying solid fuel and cleaning up dust and soot, reduction in the manual clearance of snow, and the use of aerosols. An unsympathetic view is that this growth is an index of laziness.

Table 5.6. U.K. consumption of various materials: 1

Material	*Per capita* consumption 1970 (tons)	Percentage change 1961–71
Fluorspar	0·004	+ 167
Plastics	0·03	+ 150
Salt	0·06	+ 139
Man-made fibres	0·01	+ 111
Petroleum	1·80	+ 111

Table 5.7. U.K. consumption of various materials: 2

Material	*Per capita* consumption 1970 (tons)	Percentage change 1961–71
Motor spirit	0·20	+ 75
Sand, gravel, stone	4·35	+ 70
Vehicles licensed	0·27*	+ 55
Rubber	0·007	+ 48

* Number

Source: *Annual Abstracts of Statistics*, H.M.S.O. (1972).

Table 5.8. U.K. consumption of various materials: 3

Material	Per capita consumption 1970 (tons)	Percentage change 1961–71
Fertilizers	0·03	+ 25
Cement	0·31	+ 24
Aluminium	0·01	+ 24
Sulphur and other materials for sulphuric acid	0·05	+ 21
Iron and steel	0·54	+ 18

Table 5.9. U.K. consumption of various materials: 4

Material	Per capita consumption 1970 (tons)		Percentage change 1961–71
Newsprint	0·03		+ 6
Timber	0·21*		+ 1
Lead and zinc	0·01		− 1
Soap and detergent	0·01		− 1
Food	0·57		− 2
of which: potatoes		0·10	− 2
flour		0·06	− 13
sugar		0·05	− 10
meat and fish		0·08	+ 34
Wood pulp for paper	0·05		− 7

* Cubic metres

Table 5.10. U.K. consumption of various materials: 5

Materials	Per capita consumption 1970 (tons)	Percentage change 1961–71
Copper	0·01	− 10
Bricks	108·78*	− 12
Coal	2·77	− 28
Wool	0·003	− 34
Cotton	0·003	− 40

* Number

Source: *Annual Abstracts of Statistics*, H.M.S.O. (1972).

(ii) Products that increased by 50—75 per cent over the decade, namely, vehicles licensed, motor spirit, rubber and sand, gravel and stone—representing cars on the road, miles driven, tyres used, and roadway built. It is striking that motor spirit use grew less than fuel oil use, rubber less than plastics, and road-building materials much more than house-building materials (e.g. bricks). Thus the growth in personal transport was less than that in 'convenience items'.

(iii) Products that grew by about 25 per cent. These represent the infrastructure of industrialization: aluminium, heavy chemicals like acids and alkalis, iron and steel, cement, and fertilizers.

(iv) Products that changed very little: newsprint, wood pulp and timber, lead and sinc, soap and detergents, and food. These represent traditional items that have simply held their position. Within the constant offtake of food, however, there was a 34 per cent increase in meat and fish protein, and a 10 per cent reduction in flour and sugar.

(v) Products that lost ground (minus 10 per cent to minus 40 per cent over the period); coal, copper, bricks, wool, and cotton. These were items being replaced by others requiring less labour, less investment, or both: oil for coal; plastics and aluminium for copper; cement for bricks; synthetics for natural fibres.

In summary the three main driving forces have been (and will continue to be) first, a reduction in the personal effort needed in everyday existence; secondly, much greater personal movement and mobility, of an unplanned and undisciplined kind on land, but with very great discipline in air transport operation and considerable continuing marine discipline; and thirdly, a spread of attractive food, drink, and clothing throughout a majority of those living in developed societies. The by-products have been urban congestion leading to great loss of urban amenity; a pattern of natural resource consumption contributing to a great loss of economic and political stability, and to a need for resource discipline; and the unleashing of more undisciplined demand as a result of the replacement of accepted elitism by more egalitarian payment and consumption. The prize for better surface transport and urban discipline is better cities; the prize for better general resource policy is greater economic and political stability; and the prize for less greed is better human relationships.

3. Present trends

Adaptation and innovation in response to constraint have been dominant features of man's advance ever since the Stone Age. We would be foolhardy in the extreme to attempt to divert such a rolling tide. We are certainly not seeking to suggest an entirely new type of scientific research or of social behaviour, nor do we wish to foist our own ideas of utopia on to unsuspected technologists. Rather, we aim to reinforce present trends where they are helpful, and at the same time to point out the existence of some blind alleys (and some not-so-blind ones too). It is important, therefore, to understand how the total system is changing, and whether these changes are beneficial or not, before seeking those areas in which small pattern alterations can have the greatest effect.

(a) *Extraction of natural resources*

We shall be dealing in the next chapter with capital and saving as a resource, but it is necessary to note that the mobilization of natural reserves is normally much more capital intensive than the subsequent manufacturing steps undertaken after the primary digging, processing, and transport. Not uncommonly, mining and smelting investment has to be three times the annual sales value for products such as copper, and this after quite considerable extra social capital has been provided out of public funds at the mining site. Such investment is escalating all the time (apart from the effects of inflation) because standards of pollution control and land care and restoration are being steadily raised. Moreover, ores of steadily reducing concentration are being dug and processed. Technological advance and increasing scale of operation have enabled product costs to be kept constant or even reduced in real terms. For example, the copper ores being worked about 1900 contained about 2 per cent copper: in 1972, real costs were much the same as in 1900, even though the ores processed only contained 0·5–0·6 per cent copper.[8]

As the more concentrated and readily accessible ores are used, exploration penetrates to areas that were previously thought inaccessible, such as Alaska for oil, the Canadian North for uranium or remote parts of the Pacific for copper. As an alternative, poorer material can be re-examined including spoil heaps originally rejected as worthless in and around more convenient mining areas. Use has already been made of coal tips of material rejected in South Wales for naval use around 1900, and of lead-zinc tailings. Thus, the following trends are in progress:

(i) *Exploration and workings* are moving to remoter areas, and are therefore costing relatively more.

(ii) *Concentrations of wanted materials in ores* being worked are falling, thus again increasing costs.

(iii) *Capital expenditures per unit output* are increasing because of more stringent environmental requirements, but decreasing because of scale and better technology.

(iv) *Fuel and labour unit costs* are increasing, but usages per ton of product are often decreasing (for a given ore concentration) because of better technology.

(v) *Transport* is, on average, dearer in real terms per ton delivered, although the use of bulk carriers and bulk handling is containing some of the increase.

On balance, the overall real cost trend is now strongly upward, technological improvement being unable to keep up with the opposing factors.[12]

(b) *Substitution*

The cost of recovering different materials are escalating at different rates, as are the costs of fabricating domestic and engineering items. Consequently motives for substitution are continually changing. Tables 5.11 and 5.12 give some typical calculations for alternative materials in given end uses with an indication of relative sensitivities toward energy inputs.

Table 5.11. Primary and recycle energy costs of materials

| | Energy required other than for transport and sorting (in pounds of coal to make 1 lb of material) | |
	From ore	From recycled material
Steel	1·11	0·22
Aluminium	8·32	0·17
Copper	1·98	0·11
Glass	0·36	0·36
Cement	0·33	–

Source: Faltermayer, E. Metals: The warning signals are up. *Fortune,* **86**, No. 4 (1972).

As poorer ores come into use, the volumes available normally increase, since the very high concentrations first worked usually arise from rather special geological conditions at some time in the past and in particular places. Examples of high concentrations are mercury in Spain, or gold and diamonds in South Africa. But the totally available volumes of many materials, in particular aluminium, iron, and carbon, are very large (see Table 5.1), although most of these substances are at concentrations well below those worked at present.

Table 5.12. U.K. energy requirements: fabricated products

Energy and raw material (tonnes of oil equivalent) for the manufacture of:	
1 million square metres of packaging film	
Polypropylene film	110
Cellulose film	155
1 million fertilizer sacks	
Polyethylene sacks	470
Paper sacks	700
100 km of 1-in diameter service pipe	
Polyethylene pipe	120
Copper pipe	96
Galvanized steel pipe	500
1 million 1-litre containers	
P.V.C. bottles	97
Glass bottles	230
100 km of 4-in drainage pipe with fittings	
P.V.C. pipe	360
Asbestos cement pipe	400
Cast iron pipe	1970
Clay pipe	500
Pitch fibre pipe	440

Source: I.C.I.

All present human needs can be met by materials in permanently plentiful supply, provided that the necessary energy can be found and paid for, and the plant capital provided. This apparently startling statement conflicts with much conventional wisdom, but has powerful evidence in its favour. Table 5.13 presents a rough recipe which would enable us to continue, with known materials and technologies, to lead a life fairly similar to the present. In addition, there are other abundant sources which so far have scarcely been touched. Silicon dioxide, for instance, the most plentiful substance in the earth's crust, is a basic constituent of glasses, which are remarkable versatile materials having properties that include excellent corrosion resistance. At present, however, they are used hardly at all in relation to the vast amounts available.

(c) *Recycling*

There is an important and increasing social response to resource constraint in the form of a considerable and proper interest in recycling.

There is substantial recycling already in industries using copper, steel, paper, and rubber; and many other materials, including thermoplastic and synthetic fibres, are beginning to be re-used. With a stable population consuming goods at a steady rate, one can envisage in due course reducing very markedly the demand for fresh materials toward the low level required for making up the small losses in recycling. The method would then resemble the use and recovery of a solvent in a chemical process. Recycling and reclamation also obviate part of the rather depressing task of disposing of urban waste. However, recycling consumes the scarcest of the natural resources, namely energy, which is required for the collection, sorting, and cleaning of recycled objects. Hence it by no means follows that the multiple recycled use of material A is more economic in energy than the once-through use of a less energy-intensive material B. For example, the reprocessing of metal containers may involve the remelting of steel, which may require more energy than the entire manufacture of an alternative plastic film pack, which can then be burnt (preferably usefully) with a net gain in energy usage. If the plastic can instead be re-used, there is still a gain because of the lower temperature of re-processing.

Table 5.11 shows the primary and recycle energy uptake of a series of typical materials.[13] Aluminium is of considerable interest: it is very energy-intensive for first manufacture, but reprocessing it is more economical than for most metals. Apart from energetics there are other factors that make recycling difficult. These include the following:

(i) *Inconvenience of physical form for reprocessing.* Before they can be recycled in a steel furnace hollow objects such as automobile body parts need to be compacted. Plastics and rubber require special shredding.

(ii) *Additives.* Colours and dyes are burnt away when metals are remelted but persist in plastic or fibre reprocessing; this can limit the range of end-uses. Other additives and degradation products can also cause quality variations in recycled items. The problem is especially awkward for metal alloys; and dangerous additives can sometimes be present in materials.

(iii) *Impossibility of precise quality control.* Fabricators whose costs are high suffer substantial cost increases if production is slowed or interrupted by feed or inferior quality. It is therefore advisable to direct some recycled materials towards less exacting uses in objects made by less specialized machinery.

Recycling, then, is not a simple panacea, but it should play an increasingly important part in the materials industry. It requires the following: the active co-operation of the consumer; design with recycle in mind (so that components of a material are not awkwardly united); improved extractive chemistry and metallurgy; new outlets for recycled materials that lack some of the properties of the original; and, finally, planning at industry or national level.

4. Options for the future

(a) *Authoritarian solutions*

We have already said that the nature of reserves of raw materials and energy, together with the price mechanism make the concept of 'resource exhaustion' unrealistic. The first question about future policy, then, must be 'Why fuss?' The market machine is highly adaptive, and over the past two hundred years has performed remarkably well. It has rewarded innovation, motivated substitution when a particular resource has become scarce and therefore dear, and has generally played an important part in raising living standards. By contrast, government machinery in command of technology has usually done less well, except in wartime. Models and solutions which call for highly structured resource management therefore seem to require political institutions that do not yet exist in Western democracies. Those nationalized industries that seem to be economically innovative, such as Dutch State Mines, are usually free to behave much like limited liability companies, with relatively little state intervention. Yet as we shall see, the market model, however helpful, does not seem to be sufficient.

In authoritarian countries, such as Russia, Eastern Europe, and China, state control succeeds in avoiding economic cycles, and is reasonably good at managing the process of following the Western technological advance. Where a lead has been established for a time, it has been in pursuit of a very well-defined, non-market, military-style objective, such as the placing of a satellite in orbit. The state scientific institutes lack the means for testing and responding to consumer reactions, and the integrated management of science, engineering and market sociology, which is increasingly essential for innovation, is also missing. Consequently, it is not surprising that virtually no major inventions in the domestic, civil transport, materials, electronics or processing fields have been forthcoming in Russia or Eastern Europe. Conceivably, the Chinese may do better. They have succeeded in gaining acceptance for a social system that seems to unite highly disparate elements by close intermingling: intellectuals work for part of the time at manual or military tasks. A new approach to techno-economic advance could come from such a society.

It is therefore to be expected that although authoritarian states will be better able to tackle resource problems by rationing, they will have considerable difficulty in applying highly innovative solutions. The Western democracies have the reverse situation: they will eventually produce the innovative solutions, provided that economic problems and chaos do not overwhelm the R and D machine.

This situation is not too difficult to explain and understand. Democracy permits the assembly of groupings of different capabilities that can tackle problems in new ways, and are therefore innovative. But these are the very groupings that can also very readily generate the heresies that are most dangerous to political or religious orthodoxy. They are consequently intolerable in authoritarian states, and innovation suffers. Democracy, however, also requires elections and political change at intervals that are quite short compared with the time taken for technological changes to occur and have a real effect on society. Government management of technology in a democracy is consequently liable to change inconveniently often, to be manipulated and distorted so as to catch votes, and generally to be commanded by factors that are irrelevant to techno-economic success. The forces that were essential to unseat the Nixon regime in the U.S.A. are exactly the same as the forces that inhibit good long-term resource or technology policy.

Politicians differ in the emphasis they place on strategy and tactics, but fine balances and minority governments have a greater requirement for skilled tacticians than for skilled strategists. Resource decisions have very long shadows, and very large ones. They therefore tend to be ducked by tactical politicians and muffed by tactical businessmen, because the price mechanism cannot always finance them. The normal method of borrowing money from far-sighted investors looking for long-term benefit may be inadequate, partly because politicians may castigate the patiently awaited benefits and profits as excessive in order to win votes by price-control legislation.

The option of dealing with all future resource problems by the democratic price and market mechanisms, which have hitherto been the only satisfactory foundation for innovation, therefore does not exist. The first option, therefore, is to adapt authoritarian government in order to manage the resource-rationing problems under conditions that are hostile to innovation, and to manage with no new technology. The second is to make the democratic mixed economy work better, either by introducing more freedom from political and economic stop-go into nationalized industries, or by introducing more long-term public finance, with corresponding public constraints, into the private resource industries. The third is to redesign society totally, as on the Chinese pattern.

The second option is the one we develop in this book, indicating methods of attack rather than polished answers as such. A team effort will be essential, and technologists have at least as great a part to play as politicians and economists. The problems involved are best studied by modelling, but it will not be an easy or a rapid matter to inject more political, economic, and resource stability into Western democratic mixed economies.

(b) *The use and limitations of synoptic, 'top-down', modelling*

We do not believe that existing economic models and political machinery can solve current major problems. In tackling the huge task of adaptation we have two basic procedures to help our designs: synoptic 'top-down' modelling, and synthetic 'bottom-up' modelling. As we explored briefly in Chapter 2, the first adopts a series of doctrinal or dogmatic guidelines, and then applies them to each individual situation—agriculture, industry, trade, employment—in turn. The second considers each part of the system as a separate entity, to be analysed in isolation; only after this stage can useful assembly take place, and it is not at all certain that the total structure will necessarily make sense.

The top-down doctrine may be religious or atheistic, but it must be definite and enforced, or energies required for urgent problems will be dissipated in mere theological dispute. Had resource problems been pre-eminent in the Middle Ages, the Church would have played an important part in deciding the pattern of their solution (just as the geopolitical problem of partitioning colonial rights between Spain and Portugal was settled by a papal fiat giving Spain the Western, and Portugal the Eastern hemisphere).[14] Doctrine about usury would have placed capital intensive operations under control of Jewry, which would then probably have been prevented from becoming too powerful by repression.

We already have seen two living demonstrations of Marxist top-down modelling, and it is not unexpected that the same set of century-old political writings can be held to support two procedures as different as those of Russia and China. Top-down modellers are very versatile in reading their Bibles. Christianity could be passionately held to support, on the one hand, the Borgia states and, on the other, the Quakers.

Greek political theory stood four-square and equally behind the collectivism of Creon and the individualism of Antigone. Indeed, top-down modelling is often the point of origin of the classic conflicts of right with right that lie behind the great tragedies. But the paradox inherent in such sweeping doctrines is difficult to escape: a state based on top-down modelling throughout cannot afford to be democratic and must be authoritarian if it is to be effective. We reject absolute authority in politics, and so we must severely restrict the applicability of top-down modelling. This is the position we maintain in this book.

Top-down modelling can, however, be used as a means of investigation that draws attention either to areas in which research would be quite clearly misdirected, or to fruitful ideas for pragmatic investigation. Neither the Club of Rome nor the Hudson Institute really think it likely that they will be called upon to form world governments, yet they believe that their generalized models influence actual governments, and voters who will choose those governments. The trouble about such modelling is that it can become a game that deflects able people from urgently needed action (which carries dangers to reputation and sometimes even to life and freedom) toward intellectual amusement or theological dispute. To be sure, theologians have from time to time been burnt at the stake, but only when there seemed to be a real danger that their teachings might stimulate the have-nots to seek more from the haves. It was safe to fight hammer and tongs even about the nature of the Trinity, as long as it did not stir up the Lollards or Lutherans.

Probably the best kind of top-down modelling is the variety that suggests manageable tasks that would help society if they were successful. A plaintive academic (or industrial) cry that all would be well if only people would be sensible and generalize their actions butters no parsnips. One common-land grazier can see the gain to himself by putting an extra beast on an already fully grazed common more clearly than he can understand the net loss to himself and to the whole community if everyone were to try to graze more beasts. Thus it is a very hard struggle, as successive Chancellors have found, to contain wage claims by perfectly sound arguments about inflation. By contrast, a model that creates a new preference that does more good as every new user takes it up, can move mountains. A Londoner who leaves his car at home and travels by bicycle may get to his destination more quickly and have less trouble and expense from parking. If he is prepared to brave the elements and exercise his muscles, he is contributing to overall benefit, and the more he can persuade his friends to follow his example the more pleasant and safe will be the journeys of them all.

(c) *Synthetic 'bottom-up' modelling' assembly into hierarchies*

The companion to synoptic modelling is the pragmatic assembly of sub-solutions, by viewing a series of models as a hierarchy. This has already been touched on in Chapter 4. Knowledge of what can and should be done next year can almost certainly generate further ideas about plans for the following year. Success in the modelling of traffic flows in Westminster may generate practice and principle that can be tried for London as a whole, and then—perhaps modified—for the entire U.K., taking due account of regional peculiarities. Resource models could be similarly assembled to examine substitution and price behaviour. A possible system might be shown in Fig. 5.3.

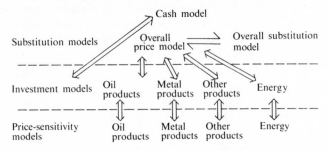

Fig. 5.3. Hierarchy of resource models.

On this basis, crude calculations would show the magnitude of the price changes that would be needed to make substitution possible in principle. Then the rate of substitutions could be determined at various rates of saving and investment, and strategies examined for taxation so as to shift consumption in any direction that is believed to be helpful, e.g. towards renewable resources, or towards investment in house insulation rather than furnaces. Recycle models and population models could be added as required, always remembering, however, that a model's main property is utility and a modeller's main aim is artistic simplicity. Only by building the models to show real situations and then testing their predictive powers by checking against performance can confidence be built up. Other hierarchies, of simpler or more complex character, will be needed for other constraint studies. Even human and organizational systems can to some extent be modelled usefully.

In the resource field, there is immense scope for the constructive exploration of sub-models, and it is not possible in this book to do more than indicate lines of attack. One key area is that of hydrocarbon and energy policy. It is useful to look at this form the viewpoint of a technologist designing a research and development programme, and then to see how the conclusions may be modified by looking at yet bigger systems, embracing the impact of energy policy on life in cities over a long period.

Most technologists had for a long time realized that supplies of oil, though cheap, were limited and that continued consumption at projected rates would soon drive prices up. However, until prices actually rose, it was difficult to justify anticipatory steps that although they were sensible would have been uneconomic in the short term. What could be done, and was done in a preliminary way, even before 1973, was to examine the changes that could be economic if a possible price change were postulated—say, a trebling in the price of crude oil. In such a scenario, it was also necessary to postulate the accompanying changes in prices of coal, nuclear fuel, capital charges, and agricultural products. It was certain that the particular patterns examined would never occur in detail, but the search was for sensitivities and for actions that were likely to be beneficial over a wide range of conditions.

The scenario could have clearly visualized an oil price made possible by concerted use of market strength by the oil producers. There was no direct technical reason why this should immediately affect prices of coal or other fuels of domestic origin in Europe or America, although it might have been foreseen that coal miners would recognize an improvement in their wage-bargaining position. The first question, then, was to model the system to answer questions concerning the possible manufacture of chemicals, including fertilizers, from coal; the relative positions of competitive vegetable-based and oil-based products such as natural and synthetic rubber, rayon, and nylon; the new preferences as between different plastics or synthetic fibres; and not least the rate at which the new preferences could have an effect.

There are, of course, many other similar questions, but these will serve to illustrate procedures.

The coal question could be easily answered. At European or American prices coal would become considerably more attractive for electricity generation, but not sufficiently so for the manufacture of chemicals such as ammonia-based fertilizers and methanol; a six-fold oil price increase from the mid-1973 level would be needed for that. This difference arose because the capital costs of ammonia or methanol plants based on coal are substantially greater that those for plants based on oil or gas. The differential between capital costs for *burning* oil or coal in power station, however, is much less.

On the other hand, in a country such as South Africa or Australia, with cheaper coal, the balance point could lie in favour of coal-based fertilizers and chemicals. It was therefore clear that technologists would need a good knowledge of geographical factors. Further, political considerations would bear on the decisions about whether capital-intensive enterprises could or should be set up in various parts of the world; such undertakings are very readily expropriated by a hostile government. Equally, balance-of-payments considerations might conceivably lead to U.K. tax policies designed to shift coal-oil preferences. Accordingly, in this pre-1973 model there was much to be said for starting a long-term interest in coal technology, but no imperative for crash action.

The effect on vegetable-based versus oil-based products was more complex. It so happened that timber and wood pulp prices had been depressed during the late 1960s by an excess of supply over demand; but in the early 1970s shortages were developing and prices were rising. It therefore appeared that the balance would not change much between natural and synthetic films and fibres, whereas natural rubber might gain an advantage that could well grow. A further point was that ethanol appeared to become economic as a source of ethylene in countries with plentiful fermentable material such as molasses. These results suggested a reinforcement in long-term research on fermentation and on natural rubber and other forestry products. But there was no immediate prospect of a big swing against polymers, and again no emergency research was required.

Preference changes within and between synthetic materials gave more clearcut answers. Products made by routes involving more steps, or less efficient ones, use more hydrocarbon (as fuel and feed) per ton of product, and the difference is large enough for development programmes to be significantly affected. For example, as Fig. 5.4 shows, sales of polyester fibre had already been growing faster than those of nylon fibre for some years, and now, in addition, had the advantage of requiring only 3·7 tons of hydrocarbon per ton of fibre, as against 6 tons of hydrocarbon for nylon. The consideration of a simple model of this kind might not always give the complete answer. For packages, the quantity to be considered is the hydrocarbon usage per unit quantity of material packaged: kilograms of meat, litres of beer, etc. These considerations can favour synthetic plastic packaging rather than cellulose or glass-based containers, despite the superficial argument that the former, being directly oil-based, might be thought to use more oil in total for their manufacture. Thin, oriented film packaging may often be better than other packs that can be recycled several times. Table 5.13 indicates other patterns for substitution.

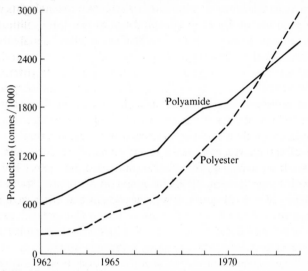

Fig. 5.4. World polyester and polyamide fibre production. (From *Bull. int. Rayon synth. Fibres Comm.*)

Table 5.13. Substitution of energy and capital-intensive materials

Purpose	Material	Replace by
Building	Bricks, cement, sand, gravel, reinforcement (steel)	(No supply problem apart from capital and energy)
Engineering	Steel and alloy steels	(No supply problem apart from capital and energy)
Electrical conduction	Copper	Aluminium
Domestic and civic piping and fittings	Stoneware	(No problem)
	Copper	Thermoplastics or aluminium

The pre-1973 use of such an assembly would not have predicted large discontinuous changes in economic and technological preferences. When the oil price increase *did* occur, however, one distribution system changed as if by a trigger mechanism. The prices of packaged tours and holidays happened to be dangerously low through fierce competiton. Cost and price increases suddenly reduced demand to the point where companies in this business suffered a further deterioration in an already unhealthy cash position. Bankruptcies occurred until the market appropriate to the higher prices—further raised by a reduced scale of operation—matched demand again.

5. Future progress

Many questions in the resource and energy field, apart from those already mentioned, will be susceptible to modelling of the right kind. As a rough indication of possibilities, the following questions would be worthy of early trial:

(i) What electricity price would be needed in order to finance the growth of electrical generating capacity from coal, or from nuclear systems, at various given rates, assuming that debt must not rise above a stated level?

(ii) How sensitive are the answers to (i) to late completion of nuclear power stations, strikes, or increases in coal-miners' pay?

(iii) What rates of oil price rise would be needed to keep a particular oil state's revenues constant for various postulated production levels (at constant percentage royalty rates)?

(iv) What costs for house insulation could be financed out of the consequent savings in extra central heating or electrical power station capacity, assuming the same house temperatures?

(v) How would various rates of increase in North Sea oil production, at various assumed oil price trends, affect the U.K. balance of payments, taking into account all capital expenditures?

The important requirement is that the researchers involved in such models should be actively involved in the consequences of their answers and in policy-making. It is on this basis that technologists can tackle constraint problems with most confidence.

Perhaps imagination is the greatest asset of any modeller. The questions above are straightforward in some ways, and it would be surprising indeed if the Department of Energy were not already tackling some, if not all, of them. To close this chapter, therefore, we first outline two synoptic, or top-down models that may just spark off more wide-ranging ideas than those we have considered so far; and secondly we paint a picture of a possible post-agricultural-revolution world, in the hope that constructive research may result.

(a) *Two new 'top-down' models*

The first of the top-down models is a 'low-temperature world' model, and the second a 'reduction of over-complexity' model. Both, as far as we know, are original. Both are intended to stimulate beneficial small-scale actions in full recognition of the impossibility of their large-scale adoption.

The 'low-temperature world' model begins with the observation that large amounts of energy are consumed in making materials and fabricating objects from them, and in running wasteful engines that heat their surroundings only too well. Further the containment of high temperatures calls for high capital expenditures for furnaces and handling equipment. Yet there are no basic human needs for high-temperature technology as such: apart from low-grade heat for personal warmth, man needs either heat or radiation only for sterilization of food and medical equipment. If, therefore, all implements and equipment could be fabricated at temperatures high enough to prevent accidental softening during normal storage and use, but no higher, there could be very large resource economies. The temperature chosen as a 'safe' level could be around 300 °C. Lower values are feasible but might give difficulty in the tropics, and 300 °C is low enough to provide the inventive stimulus that is the aim. The basic assumption, then, is that after a period in which economic penalties were used to discourage the use of high temperatures it would be made illegal to generate or use temperatures about 300 °C. This, of course, rules out the use of fire and steel, which sounds drastic enough as a starting-point.

Medical sterilization could then be conducted either by radiation or by autoclave operations, and so in theory could food preparation. The human alimentary canal has, however, become accustomed to cooked food (even though it has probably overdone the reduction of roughage in diet). Moreover, cooking and gastronomy are valid art-forms, and man does not live by hamburgers or fish fingers alone. With a top limit of 300 °C, and the use of microwave equipment as an occasional addition to the casserole and pressure cooker, cooks and chefs would still have plenty of scope. And there might conceivably be a notable reduction in cancer because of a reduction in the intake of pyrolyically generated polycyclic hydrocarbons and other aromatics. No basic inventions, therefore would be needed for medicine or nutrition.

Metals would have to be smelted electrolytically, and deposited and shaped either in the same way or by the controlled decomposition of labile organic compounds such as carbonyls. Electric motors would have to be used for all work and mechanical locomotion, and electricity would have to be generated by fuel cells, which would therefore need basic improvement. There would be a great incentive to use organic polymers, including timber, for construction and for the support of metals wherever possible. Indeed, probably the only economic use for metals would be for the conduction of heat and electricity, and the former would greatly fall in importance in a low-temperature technology.

The full economic benefit from such a scheme might not be achieved unless it were pursued as a total system, but on the other hand, there might be some tasks that could be done so much more cheaply at high temperatures that it would be absurd to avoid them. Almost certainly, however, areas would be discovered for co-operative invention and the development of new procedures. There is, for example, a tendency for large-scale organic process chemistry to move into this lower temperature range—but under conditions appropriate for containment in mild steel. It would be an interesting challenge to devise processes and catalysts that could be entirely conducted in plastic composites, or to invent building systems based on sophisticated thixotropic and hard-setting muds made without firing any of the materials. As invention progressed, it would be possible to identify the energy price at which each technology would become economic, and then to devise entrance strategies.

The second top-down model of this kind, the reduction of over-complexity, is similar but more general. It will be seen that the low-temperature world model arises from the 'disinvention' of technology that is at present useful but that may become wasteful and diversionary. There are many other targets for 'disinvention' by creating better inventions for the new conditions. The processes for making garments from fibres were logical enough when cotton, wool, and flax were the only alternatives to skins and fleeces for clothing, but are illogical now that there are polymers that can be melted, oriented, and fibrillated by various kinds of pulling and teasing. The blowing of garments, or spraying them on to tailor's dummies, could well be a fruitful area for exploration. Disinvention in the medical field needs much more careful thought, for it may not be entirely sensible to eliminate diseases from which people can die comfortably at (say) 80, in order that they can survive under conditions of senile dementia, or alternatively die from something particularly unpleasant at (say) 82.

The subtle purpose of 'disinvention' models would be to motivate the processes of irrational invention, such as have driven most important advances in the past. It is essential to avoid falling into the trap of saying what positively should be invented; this merely deprives the process of its most creative element, by making it rational and therefore probably dull and ordinary.

(b) *Agricultural models: the rural society*

One of the most important social requirements in developing countries is to avoid increasing the already unacceptable number of urban poor. If agricultural development along present patterns proceeds without industrial development, dispossessed farmers are forced to emigrate to already impoverished and overpopulated towns. Production of more food tends to exacerbate the very urban poverty it is meant to cure, and the provision of bread without jobs does not lead to a stable society. Worst of all, rural poverty makes population control nearly impossible, for the only provision for old age becomes the rearing of strong children. The solution at the individual level prevents a solution at the community level.

On present practices and perceptions, it is conventional to say that rural and urban development must go in step. However, now that advanced agricultural and dietary procedures are available, there may be a better line of advance in seeking development through a coupling of labour-intensive farming with high productivity per unit area. Such conditions already exist in Japan, Taiwan, and the Nile Delta, and there is a possible reinforcement available through new development in manufacture of materials and fuel from the land.

Production of fuel (such as ethanol) and polymers (such as polythene made from ethanol) could with advantage be micro-urban or macro-rural. This would avoid the double transport of agricultural produce to big central processing plants, and of industrial products out to large numbers of consumers. A neighbourhood 'agricultural simplex', producing some tens of thousands of tons of fuel and polymer annually, could be fed from a catchment area of some miles radius by farm vehicles. And although this would nominally reduce the land available for food, it would generate cash and capital locally (which it is wanted), build up industries responsive to local demand, and help to underpin much-needed rural education. But it is likely to come soon only in rather limited areas, and it will be most difficult to achieve where populations are largest, as in India.

One unquantified threat is at present outside the range of technology; his is climate fluctuation or secular deterioration. There are current predictions of poorer weather; severe sub-tropical drought and extension of desert are already evident. This may be the area where new technology is most needed for food production, but there is as yet little work in progress.

6. Conclusions

Energy and raw material constraints are not new and are not catastrophic. The study of ways and means for making better use of energy sources and of materials will exercise the traditional abilities of technologists for many years to come. This approach must lead towards the replacement of processes that depend on fossil fuel reserves by others that are conservative and depend on current or recently arrived solar energy, or are highly innovative and depend on nuclear fusion. Virtually all other supply problems, including those for food, can be solved by the availability of enough energy. Food supplies can be maintained and increased by a rich variety of methods, but these will together call for a second agricultural revolution, the result of which will be the co-production of food, portable fuel, and organic chemical raw materials for plastics, foams, films and fibres.

This said, the reason for the current severity of resource and energy problems is not exhaustion; it is adjustment in human behaviour and organization in response to the prospect of supply difficulties. This adjustment is chaotic and disorderly, and must be studied and understood by technologists if they are to devise good priority systems for both short-term and long-term solutions. Hanging over the entire system is the need to stop population growth soon; this is now more a behavioural than a technical problem. Also, response to the shortage of capital, resulting partly from loss of confidence in the value of saving, sets new technological constraints. But given sufficient imagination and communication, we believe firmly that the problems are soluble.

6

Capital and finance

1. The state of the world

(a) *General*

We have already discussed the connection between saving (i.e. postponement of consumption) and technology. In effect, technology is the principal means whereby savings are converted into real benefits. Clearly, society will continue to purchase the benefits only if they are seen to justify the associated saving. It is just as important for technologists to recognize that their subject is no more than a means to socially desired ends as it is for financiers and economists to remember that money and its manipulation are means for socially desired exchange, and never ends in themselves. If in the long term society ceases altogether to desire certain technologically generated goods, let alone the further development of such goods, then technology can and should be correspondingly reduced in scope.[1] The processes of generating and allocating investment capital are very important means by which society in effect indicates its decisions to the technologist. The technologist must therefore understand the financial constraints on him if he is to act on these messages early enough.

The very concept of possible plateaux or contractions in technology would have shocked many people a generation ago.[2] At that time, every achievement that permitted faster or more distant travel, longer bridges, taller buildings, bigger ships, or new household devices seemed at the same time to bring associated social benefits. There still exists a substantial school of thought that says 'faster is better'. Any view contrary to this, even when held by a majority, is countered by the statement that it is only an outbreak of temporary myopia and that people have often at first been doubtful about 'progress'. Now it is certainly true that opinion is fickle, and that there are many occasions when an initially unwanted development has, in the end, won universal acclaim. There are therefore powerful arguments for avoiding instant action based on short-term judgements about the likely benefits and penalties of new technology.

This said, we can now witnessing much more hesitation about investment in new technology.[3] Society is voting with its money, as it were, and is giving much more careful consideration to the savings and the postponement of consumption that will be required. There are three basic reasons for this. First inflation and lack of confidence about the future erode profits from manufacture and thus redirect savings (which, rather surprisingly, rose in the United Kingdom during 1974—5) toward non-productive assets that may rise in value.* Secondly, the 'entrance fee' for a new technology is generally very much bigger than it was.[4] Thirdly, society now has a more critical attitude towards novelty, for it already has so many technological innovations in process of social acceptance and dissemination. Each of these reasons separately would cause the technologist some disquiet; taken together they radically alter the way in which he must seek and reward financial support.

Far more cash (even in real terms) has now to be put at risk before sales of an economic and acceptable airliner can cover the manufacturing cost and begin to pay back the development cost. The entrance fee for the Boeing 747 jumbo jet may have been around $1000m, compared with a very small fraction of this for the DC3 of the late 1930s. The figure for a completely new general-purpose plastic or textile fibre might now be in excess of $100 m, compared with perhaps 10 per cent of this sum needed to launch the economic manufacture of nylon in the 1930s. The costs of these developments are met by savings which expected to be rewarded; the reward for the pre-war successes was substantial and was not too long delayed, so that the savings could be repaid or reinvested in some other enterprise.

Technologists can respond to capital constraints in three ways. First, they can be mindlessly militant in support of the proposition that 'faster is better'. This attitude is usually accompanied by a social model in which far-sighted and altruistic scientists and technologists battle doughtily against myopic and restrictive treasuries and treasurers. All money wrested from the system is held to be justified since, like the

* Comments really refer to insurance companies and pensions funds only. Insurance premium payments and pensions fund contributions generate substantial saving by institutions, whose managers, however, are constrained to play safe and conventionally, and not to invest in technology as individuals once did. In recent years they have built many office blocks, and even lent money at fixed interest to local authorities, thus financing various kinds of civic amenity.

bread cast upon the waters, after many days it will return. It may also be held, as in the case of other priestly religions, that science cannot be rigorously explained and (a little learning being a dangerous thing) there is neither hope for, nor need of, comprehension of science by the non-scientist. On this view, popularization, being imprecise, is dangerous. And, it is added, if the private sector of the economy can or will no longer pay, the project should be moved into the public sector (or vice versa).

The second attitude is submissive. Others, though ignorant hold the purse-strings, and although it is unfortunate that they restrict resources it is best to do what is recommended, for this earns security, respect, and honour.

The third attitude is participative. It rests on an assumption that science and technology must, in the end, share perceptions, objectives, and priority systems with the rest of society;[5] and that more is gained than lost by mutual understanding. Those adopting this posture face an initial risk: society may, prior to such mutual understanding, be supporting the imperial splendour of their particular discipline or organization because no small boy has yet asked why the emperor has no clothes. Under these circumstances, their candour may be regarded as a treasonable disservice to science by their colleagues who see themsleves as dependent on imperial credibility. The participative approach has a source of strength that has always been used to some extent, and that could now be reinforced. This is the pursuit of much greater social value for the capital already invested in innovations, by dedicated follow-up and improvement. However great the initial benefits, it always turns out that further inputs of good science, bright ideas, or perceptive social thinking can improve new technology without the massive entrance fees that are characteristic of further basic innovation; sometimes, major new utility can be obtained with hardly any new capital at all. The inventiveness in such improvement is often just as great as that in the original innovation, although it is rarely as spectacular, often less esteemed, and sometimes almost held in contempt, especially by academic scientists. The quietening and stretching of aircraft; the increasing of reactor throughput by better catalysts; the faster processing of fibres; double-cropping or intercropping to increase food yields per acre; all these simply make better use of capital assets already in existence. Sometimes, because they are less spectacular, they do not come about until the absence or shortage of new capital leaves technologists with no other options. So far, innovation and consolidation have not been jointly planned, and their interdependence has not been coherently studied or modelled. It is certainly true that improvement and consolidation crucially require a minimum of risk-taking in 'forward leaps' that may be expensive and must sometimes involve disappointment or failure, but good planning and greater esteem for imaginative, value-analysed adaptation might well reduce the required minimum, and thus enable technologists to do more with such capital inputs and entry fees as are available. We shall return to this point.

It will be clear by now that the exposure of options and the sketching of models that have been attempted in this and other chapters assume adoption of the third, candid, participative, and persuasive policy. Our approach is based on clear statements about what has been, is being, and will be done, together with realistic assessements of actual or likely achievement. The technologist who makes unusually modest promises, but always (on average) delivers roughly what he predicts, wins in the end.[6] But fundamental to our approach is the notion that the technologist cannot even make the promises unless he fully understands the framework within which he is working. We shall therefore now examine in more detail the reasons for greater capital constraints.

(b) *The decreased willingness to invest*

Saving is intended to bring about a more secure existence. This encompasses the provision of better medicine, the replacement of unreliable hunted or scavenged food with reliable cultivated food, and the construction of devices that supersede toil. The desirability of these benefits can be crudely quantified in the form of an interest rate that rewards savings invested in them. There is a general and intuitive concept that in a steady, non-inflating economy interest of 3 to 4 per cent per year is 'natural and just'. This happens to be the interest rate that will transform savings of about 10 per cent of income into a satisfactory pension for the whole of life after retirement at 60 to 65, but nature and justice in fact have little to do with this concept. The reality is that for a century or more the prospect of these moderate but positive and secure returns was attractive enough to ensure an adequate investment in technological change. Technological advance could proceed at a rate which was neither too slow for the reasonable dissemination of benefits nor too fast for the introduction of changed social behaviour. If people feel instinctively that such rewards will *not* be forthcoming, despite well-laid plans, then they will not see the point of saving in the form of money. They will either consume everything earned, or save in the form of goods so as to ensure at least some form of benefit (though not the flexibility that money can bring). If, at world valuations, a country or an economy produces less than it consumes in goods and services, there is inflation and the currency progressively buys less. With a fall of 1 to 3 per cent a year in the value of money, saving can be maintained only if the interest rate goes up above the 'natural' 3 to 4 per cent level, so as to maintain the real value of the investment *and* pay the 'natural' interest. If, however, inflation rises towards 20 per cent a year (as in the U.K. in 1974), then, because one can never know what the true inflation rate is, there may be uncertainty whether savings are earning natural interest *and* making up for inflation, or indeed whether they are simply losing their value. Even if savings do not decline, they will tend to pursue high short-term interest rates rather than manufacturing profits. These are held down by the increasing costs of new plant and of necessary stock holding (see below).

This argument about the effect of inflation is quite independent of the causes of inflation, and for our present purposes it is enough to say:

technology depends on investment;

private investment depends on confidence;

confidence depends on good hopes of reasonable rewards from investment.

This simple model leaves the question of whether in an inflating economy public investment from higher taxation can take the place of private investment. In principle it can, given resolute government, but there are two difficulties. The first of these is the temptation to cut taxes and then to cut investment as the least immediately important matter, in order to win an election. Public investment can thus be dangerously susceptible to electoral pressures. The second is the indirectness of the contact between the taxpayer or pension fund contributor, as provider of the investment, and the investment that is eventually made. Instead of going toward the investment that will best safeguard the finance the investor's pension, or the hospital and educational facilities that he or his children will want, the enforced investment may be used to prop up bad projects that happen to employ a lot of voters in a particular minister's marginal constituency, or to provide Scotland and Wales with badly sited steel plants that are uneconomically small, as a sop to national political lobbies.

Underlying the world-wide inflation of 1973–5, and the decreased personal willingness to invest in technology, is a curious reality: that the global margin of cash not actually spent on consumption has not diminished and may well have increased.[7] It has, however, been heavily concentrated in two places: public treasuries, national and local (through increased taxation and borrowing), and primary producing countries, especially oil states. Governments, as we have seen, are not very expert at investing beneficially in new technology, and do not provide the technologist with the discipline, guidance, and encouragement that he has had from the market mechanism. Oil states are naturally doubtful about investing in factories built in the developed industrial countries from which the money has come, partly because of fear of expropriation and partly because of inflation. The available margin is consequently either not being invested or is being invested rather inefficiently in support of short-term political ends.

Different countries are adapting to this new overall situation at different rates, and some Western countries, notably the U.S.A., are bringing inflation under control so that the private savings and industrial investment mechanism is recovering; not so in the U.K. What seems likely, however, is that overall industrial expansion will remain at a rate that is far enough below that of the 1960s for industrial employment to continue to fall, increased productivity more than providing for the market expansion. If so, world unemployment will remain high and perhaps grow further, until the service sectors or agriculture take up the slack. Attempts, either by unions or government, to maintain industrial jobs by restrictive methods will make their industries first uncompetitive and eventually bankrupt, unless they can be sustained by subsidies from elsewhere in the economy. Technologists in the U.K. will face special problems in this area.

A great deal of work is being done to measure the effect of inflation on industrial cash margins available for ploughing back into the business. Proper treatment of this subject calls for comparison of complete models of money flow that include cases where money loses value at various rates (and 'inflation accounting' is currently an important topic.)[8] But it is easy to see, in qualitative terms, the sort of factors that come into play. For example, all industry depends on stocks of materials, work in progress, and finished goods that have not yet been paid for: during inflation, this 'working capital' is increasing in money value very fast. The extra money has to be provided from trading margins, or borrowed, and either process erodes the ability to invest in equipment. As a second example, more money has to be set aside for a given new plant, because costs of construction will rise during building: since the plant is not earning until it is finished, there is a further call on the available cash, which cannot be met by increased prices for finished goods. Hence, the technologist has to face reduced internal funds available for capital assets, as well as reduced external borrowing ability.

(c) *The increase in entrance fees* [9]

When research was less expensive than now, and there were many craft-based technologies for science to invade, reform, and improve, the advance that could be made by a new invention was often great enough for manufacture to be economic in a cheap improvised plant. Consequently, research and development were quite reasonably conducted as 'overhead' activities. They were not thought of as major calls on savings or investment until something that had started in a small, but self-supporting, way was beginning to command markets big enough to call for more specialized, larger plants. These could be on a scale to cope with a reasonably predictable demand, and thus could be fully utilized almost as soon as they were completed. Only at this stage, when there was an established business, was serious attention given to calls on capital.

A second favourable factor was that research or development successes forty years ago, in the early years of an industrial thrust newly based on science, were more frequent (though often smaller in scope) than today. Motor-cars were made in more shapes and sizes, for instance, and various inventions were tried out at modest cost and relatively low risk: hydraulic torque converters, torsion-bar suspension, drives and transmissions of various conformations, innovative valve assemblies, not to mention many improvements to comfort in seating and moving. Aircraft were similar: all-metal fuel-carrying wings, retractable undercarriages, widely varying wing arrangements, helicopters, autogyros, pick-a-back planes, large flying-boats, amphibians, all were tried in quick succession. In retrospect, it can be seen that this variety owed its existence to the fact that road and air transport being new and offering new amenity, commanded prices in terms of, say, pounds of beef that were much higher than those expected today. Chemical inventions were also multifarious and cheap, and expenditure on assessment of safety, though significant in aircraft, was not large elsewhere. New technology was simply not thought of as particularly hazardous.

However, once a new science invasion had generated a fair range of successful products that worked, then, as we saw in Chapter 3, the drive for cheapness and wider sales followed. The advent of the automobile assembly line immediately confined the entrepreneurial, batch-scale workshop producer to the speciality motor-car markets; and the addition of component supply systems, special paint booths and baths, and other high-productivity developments made tooling-up and development of a new model a multi-million-pound affair, even before a single car was sold. Continuous rather than batch production did the same for chemicals: unit costs in a fully occupied large plant came down dramatically. New products and processes had to be not only better but cheaper than very good existing ones whose particular characteristics were familiar to the market. Once rubber, PVC, polythene, and nylon were well-established, for instance, a new polymer invention had formidable consumer conservatism and habit to overcome, even without the prospect of counteracting very low prices.

Innovators also had unpleasant decisions to take about scale of manufacture. All single-stream processes where the output can be lifted by increasing the size of the stream enjoy economies of scale. In petrochemicals, for example, capital charges per unit output can come down by about the inverse cube root of the output rate, so that they are halved for an eight-fold increase of scale. Manning, instrumentation, and energy costs can also be reduced, so that all unit costs other than those for raw materials (which in any case are often only 20 to 40 per cent of the total costs) decrease as an existing product becomes better established and is made in greater quantity. Fig. 6.1 shows the way in which polyethylene costs fell over the period 1950–65, and the way in which the output of manufacture changed.

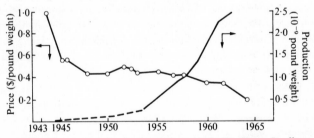

Fig. 6.1. Production and price of polyethylene 1950–65. (From Bradbury 1969).

A new invention which is intended to perform some tasks better than an established product, like steel or polythene, thus faces a very unfavourable cost differential. Unless it has properties that offer very substantial advantages and are directly useful, the price for it will be determined by the price of the existing product. But if a small plant is constructed then the costs will be too high; whereas if a bigger plant is built that will be economic when the market has been developed to the point where plant capacity is full, then the delay while the plant is partly idle and capital under-utilized may extend to several years. Either course will therefore lead to initial heavy losses. In effect, any latecomer into a business area has to spend money simply on catching up, and this forms the bulk of the 'entrance fee' phenomenon.

Schematically, the cash flow diagrams for typical polymers invented in 1935, 1945, and 1955 might appear as in Fig. 6.2. In 1935, quite a small plant would enable product A, invented for a new market, to be sold profitably. But ten years later A would be developed, command a substantial user industry, and be made on a much larger scale; so that the cash required to get a newcomer B off the ground in 1945 would have to provide for additional research to meet the more exacting needs, a larger plant, and a bigger market development cost, as well as the expense of uneconomic manufacture in a partly loaded plant in the early years. By the time it broke even, much more cash would have been required for B than A. After a further few years of growth of scale for A and B, a newcomer C would face an even more formidable struggle and an even higher entrance fee, to cover the build-up in an even bigger first plant. Further, these big first plants can and do go severely wrong, for they have been built without the advantages of experience on a smaller scale enjoyed by the earlier products. Breakdown or modification of such units is expensive and adds further to the entrance fee burden.

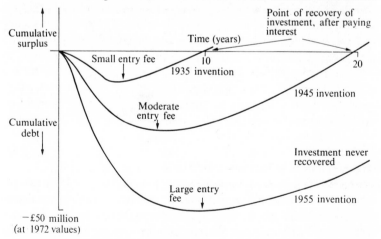

Fig. 6.2. Cash flows of typical polymers invented in 1935, 1945, and 1955.

Growing product sophistication and market demands can introduce yet further additions to entry fees. One such is toxicology, which is crucial for pharmaceuticals and crop-protection products, and is of growing importance even in the plastics field. The economics of pharmaceutical products, which are made on a small scale and sold in small doses, are more importantly affected by the cost of proving in laboratory tests that they work, proving them harmless, and then establishing in medicine that they work safely, than by production costs. Toxicological costs are not high for products (e.g. penicillin) that kill invading microorganisms but have no appreciable effect on mammalian physiology. But products designed to act on some parts of the central nervous system (or on cardiovascular action) are more likely to do other, unwanted, things as well. The resultant studies are prolonged and expensive, and contribute to a much higher overall failure rate than formerly. Hence very much more exploratory research, pharmacology, toxicology, and clinical trials are needed per successful drug. Fig. 6.3 shows some actual comparisons to illustrate the escalation of entrance fees.

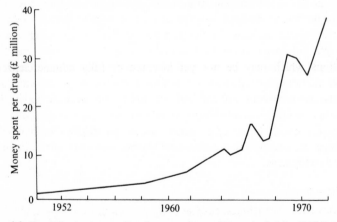

Fig. 6.3. Escalating entrance fees for new drugs. (From A. Spinks (1974) *I.C.I. Magazine*, 52 (411), 122–8.)

There can, of course, still be successes, but they must be suitably chosen innovations that avoid the danger of mere imitation of existing products. They must be backed by determined, large and competent developers, although the innovation itself may of course be better made by a smaller unit which then requires a large partner for the development stage. These successes must also, however, be good enough to pay for the failures that swallow a lot of money before being abandoned.

All in all, any given technological advance will now require more post-poned consumption than it would have done fifty years ago, and will therefore need to be more acceptable in social terms.

(d) *The declining value of novelty*

It is crucial for technologists to realize that society's rather blasé attitude to technology is more likely to become selective than general. Concern about population size, health, and food supplies may produce a greater willingness to save and postpone consumption in order to improve hospitals, birth control, and farming; but to save for innovations in cars, clothes, or hardware generally may in due course correspondingly seem less attractive. As education is more widely disseminated, this trend may be further reinforced.

A modern family has its motor-car in which to visit and explore, its TV set for a different kind of exploration (or, less sympathetically, harmless narcosis), better medicine, and more attractive and easily prepared food. It may, if it lives in a large city, be having to pay high prices in the shape of congestion, noise, and aggression that it is only just beginning to recognize as a concomitant penalty for technology. Thus the technologically equipped family has enough existing choice of new joys (or new toys) to be less insistent about the need for further additions which may be not yet invented or fully commercialized. Desire for a second car, or a more mechanical kitchen or laundry just like the one next door, can compete very effectively with the wares of a company peddling videophones, video-recorders, or (in due course) three-dimensional holographic TV. If people are (rightly or wrongly) unconcerned about new inventions, it will be more difficult to finance their development.

2. Present preferences

We drew attention in the last chapter to the economic growth-patterns that laid great emphasis on labour saving, personal mobility, and better food. The promotion of continuing investment in helpful technology offers some good possibilities that do not conflict crucially with preferences. Clearly, innovation which reduces the amount of capital required in a technology should be encouraged. There must, however, be doubts about any proposals for less capital-intensive technology that makes greater demands on increasingly scarce and expensive materials or energy resources.

One area of capital-intensive activity is the textile industry which, starting from natural fibrous materials, has built up an increasing requirement for sophisticated machinery to transform one-dimensional fibres into two-dimensional woven or knitted fabrics, and then into three-dimensional shells (garments). Typically, £1 of investment in making finished fibre from oil calls for £4 of downstream investment in weaving or knitting, decorating and colouring, and making-up. This investment could be replaced by a much smaller investment in non-woven fabric and garment technology. Progress along these lines is already considerable. One could also visualize the spraying of garments by mixing foam, or other suitable chemicals, in a stream directed at a 'tailor's dummy' with zip fasteners already in place, and with patterns produced by the spraying procedure. Before proceeding too far with this model it is advisable to ask what purpose garments serve, and to what extent temperature and humidity control play a part. Recently, temperature and humidity in buildings and vehicles have sometimes been controlled at levels that made warm clothing almost optional, if not redundant. This, of course, may well change as fuel becomes dearer, and the sweater may return to its former status as a provider of warmth. But garments may really be more important in support of role-playing— a harmless, delightful, beneficial and usually non-resource-intensive activity. If, as a result of asking the question 'Who am I to-day?', people were to dress accordingly and to spend more time sitting together and talking so as to establish admiration for Joan of Arc, Charlotte Corday, Brunnhilde, Mrs. Pankhurst, Sir Francis Drake, Oliver Cromwell, Oscar Wilde, or even Alf Garnett, then capital and resources would be saved because this activity might reduce the frenetic desire to be pro-pelled at high speed in a heavy, sophisticated, fuel-hungry, steel container.

Another area of social preference that can guide technology in its efforts to reduce demands for saving is the development of medical technology which reduces admissions to hospitals. Any time spent in hospitals, whether in public or private beds, is demanding of both capital and resources. It is, for example, important to develop pharma-cology and therapy that avoid the need for a protracted stay in a psychiatric or geriatric hospital, either of which requires a large supply of compassionate and patient people as well as capital and fuel. It is also important not to make a decisively better job of medical technology for physical ailments than for mental decay. If the technologists are not careful it will become the common lot of man to live up to and beyond 100, only to suffer from senile dementia for the last ten years. Such an imbalance would create impossibly high capital requirements for geriatric hospitals, as well as indignity and suffering for everyone. Medical imperatives accordingly need watching.

Considerations of this kind must greatly influence technologists in their search for routes round important constraints. Clearly there are possible ways forward, and beneficial innovation need not be stifled by the capital factors already explored. But technologists who think of tomorrow only in terms of a bigger and better yesterday will miss the main opportunities.

3. The options

(a) *Continuing as at present*

Of all the constraints we are considering in these chapters, it is perhaps in the field of capital expenditure that this option seems most evidently and immediately unrealistic. Inflation has not only accompanied and intensified unwillingness to invest, but has also added further to capital requirements by increasing the value of work in progress and therefore increasing the requirement for working capital. Human behaviour, the most basic of the factors imposing constraint, has put in a decisive appearance.

The traditional capitalist view of investment is that profit margins are used first to provide depreciation (which in the absence of inflation maintains assets and replaces, but does not expand, them) and taxes, and that most of the rest is distributed as dividends. These are then reinvested entrepreneurially, through the stock exchanges, in response to assessment of opportunity. A variant of this has been the appearance of banks as big investors and insistent dividend collectors and redistributors. Indeed, in Germany and Japan the banks play a big part in the pattern of national investment. The concentration of considerable private savings in insurance, pension, and unit funds has led to further institutional investment so that most dividend payments are now reinvested in one way or another.

Alongside this trend, particularly in Britain and the U.S.A., there has developed a very considerable retention of industrial profits over and above the payment of dividends. In effect this represents internal ploughback of money beyond that needed to replace assets. To be sure, inflation has made it necessary to depreciate assets over less than their real life, or to put money to reserve and then to make free issues to recognize this ploughback. Considerable industrial autonomy has resulted from this practice, since it is after all a form of zero redistribution. Harm may be done if this ploughback results in too heavy a reinvestment in traditional industries that can manage with less. It need not be disadvantageous, however, if the companies concerned are both innovative and diverse enough to take sufficient money from older, slower-growing activities and divert it to new and promising fields. It is especially advantageous for this redistribution to have an international component, for the national interest is best served by using technology where it can do most for national objectives. An important contribution to exports, and to the balance of payments generally, is made by international business operations that ensure adequate marketing—covering both local manufacture and exports from the U.K.—in all major countries and economies.

Ploughback will be even more encouraged by the disarray into which inflation has thrown the stock markets. With a shortage of immediate available cash, interest rates have climbed and prospects for continuing expansion on the pattern of the past half-century have turned sour. Investment on the Stock Exchange has traditionally assumed that growth and diversification of manufacture would more than keep pace with inflation; moderate internal finance was countenanced and as a result modest dividends were paid, well below current rates. Fig. 6.4 shows this relationship. Confidence in market expansion is now poor,

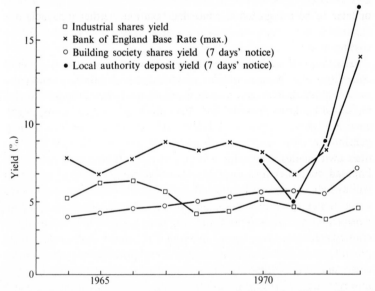

Fig. 6.4. Relationship between interest rates and dividends. (From *Financial statistics,* Central Statistical Office, H.M.S.O. 1975.)

and with statutory dividend limitation, share prices have not kept pace with inflation. In 1974, shares became cheap enough for dividend yields to move quite near to current (high) interest rates. This made it so onerous to raise capital by equity issue that investment could come only from ploughback or expensive fixed-interest borrowing. Since then, share prices have partly recovered, to the point where (in 1975) there were some rights issues, but partly by insurance and commercial companies, as opposed to manufacturing industry. There seems no prospect of an early return to the situation where expansion or innovation could be mainly financed by equity issue.

In effect, saving and investment have been decoupled; savings in U.K. are higher than ever (around 14 per cent of GNP), partly because of high compulsory saving through big mortgage and associated insurance premium payments and partly through traditional cautious reactions to recession. Yet the savings go mainly into fixed interest borrowing, especially by local authorities. This money buys non-productive (though highly desirable) amenities such as roads, sports centres, and housing, with the implicit assumption that the interest will be provided by tax-payers or ratepayers, who of course are the same people who provided the savings! It is thus reasonable to suggest that real saving is considerably less than apparent saving, and that the 'lost differential' has sooner or later to be recognized as concealed taxation, neither repayable nor rewardable with pensions or insurance payments of the expected purchasing power.

The resultant position is delicate and sensitive; any outbreak of debt collecting, such as that which follows a bank or insurance company failure, must disclose that money is not where it was thought to be. A big urban bankruptcy would have the same effect, which is why threatened insolvency in New York led the teachers to lend $150m of their pension fund to the city, partly to ensure the continuing payment of their own salaries. Nor is the way out obvious. If stronger controls are imposed on prices than on wages increases, margins (and ploughback) will dwindle. If controls are eased, inflation will accelerate and call for more money to finance stock appreciation. A general recognition that personal consumption is too high, and that it has to fall, increases unemployment. All these factors mean that the technologist has, for the present, to do better with the plants, equipment, and tools that already exist, and to live with very selective expansion of investment. There is, however, a possible compensation. Such reduction of consumption and investment has to come within a few decades because of reduction in the reserves and accessibility of convenient fuels and raw materials. In the absence of malfunction in the social machine, quite sudden outbursts of serious rivalry for natural resources could well occur. Thus, if economic muddle can continue without disaster and system failure, and if non-resource-intensive and socially esteemed jobs (e.g. service activities) can spread rapidly, the energy that might have gone into war for unnecessary rights to consume may be absorbed in conflict with social inefficiency, and so be dissipated. It might help if technologists remained clear-headed and selected tasks that fitted with the need for employment and increasing difficulty of access to resources. Insulation and low-energy recirculation would become the order of the day, rather than the pursuit of greater comfort and amenity along previous pathways. The need for sophistication and level-headedness would be higher than ever before, but again in the new direction of the understanding of people. Preservation of freedom of speech and behaviour might require considerable sacrifice of freedom to increase consumption.

(b) *Synoptic 'top-down' modelling*

Factors affecting the distribution of investment between public and private sectors, between different industries, and between different geographical regions, have been of major political interest since Western governments became involved in the economy. Consequently, qualitative and intuitive models of investment patterns have been tried out for a long time.

Models and perceptions, however, are diverse and unreconciled. An example of this is to be found in the hopes and realities of selective government incentives for investment in particular regions or industries. The simple view of the politician is that an increase in taxation, with an equal increase in selective grants, should maintain investment at a constant level but direct it in chosen ways. The reality is that such a policy tends to diminish investment because businessmen respond more readily to tax increases (which experience shows to be durable) than to special grants (which experience shows to be fickle and unreliable unless one can be quick).

As mentioned already, the confidence essential for investment, which has always cycled disconcertingly, now seems to be in longer-term decline in Western democracies. It is hard to say how far this will go, but it has already gone far enough to invalidate most the the infant investment models of recent years. Some of these were designed to produce counter-cyclical behaviour, and some the avoidance of inconveniently large regional concentrations of investment. All employed tax devices as control and guidance systems, and the practical results have not been good or encouraging in Britain. Better progress has been made in the U.S.A., West Germany, and Japan. Rather surprisingly, authoritarian models in communist societies have shown the same lack of success, since human factors—personal ambition, local deception, dissidence—cause five-year-plans to fail even when there is an imposed top-down model. All these observations combine to argue the need for technologists to have a sensitive awareness of human behaviour and organization, and to trust synoptic models only when they seem to accord with human perceptions. It may well be some time before reliable and useful integrative thinking about savings and investment can be carried very far.

One of the difficulties lies in the unrecognized divisibility and separateness of money systems. Large industries are continuing to operate on investment criteria that are not shared by the general public and the stock exchanges (which tend to amplify and exaggerate both public malaise and public confidence). Having in mind longer-term views of investment needs, industrial ploughback takes money that could bring better immediate returns on short-term loan. Industry is therefore running a money cycle with a different time-horizon from that on which spare High Street cash is invested. It is not surprising that there is at present little or no cash flowing voluntarily, apart from prices paid for goods, into the industrial cycle from the High Street (whether from the consumer himself or from his bank or insurance company). The two currencies are operationally distinct, and are only partly convertible. Similarly, the government cash cycle is used according to rules that are not shared by consumers as spenders (though shared by a majority of them as voters). Citizens would not invest in the government cash cycle voluntarily; they therefore have to be taxed, or invited to contribute to National Savings at a negative real interest rate. The three main currencies might as well be zlotys, dollars, and Kenyan shillings, and for many purposes it is a definite handicap that the three are in fact all circulating in the same territory, and are represented by indistinguishable pieces of paper.

Conventional economic models normally treat these currencies as indistinguishable, which they are not. Until models are nearer the reality, serious confusion will continue. Fig. 6.5 exposes the main cash links between the three currencies, but is drastically over-simplified and omits any indication of flows within each currency system.

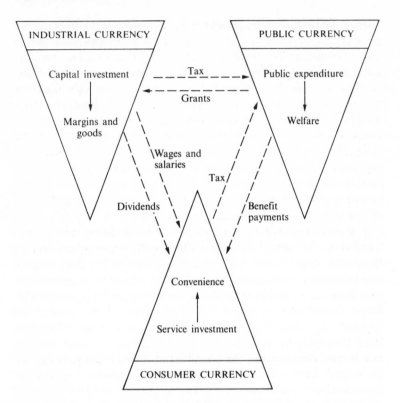

Fig. 6.5. Currency systems.

(c) *Synthetic 'bottom-up' modelling*

Meanwhile, the smaller models used to decide between different investment projects within the public and private systems are also in difficulty. Private sector models have always been easier to construct and
use because of the common acceptance of return on investment as a
unifying criterion. Even then, there have always been difficulties because
of differences in emphasis on short- and long-term return. Models such
as that of discounted cash flow can be used to take account of the
timing of monetary outlays and benefits (see Appendix). Such an
arbitrary 'snapshot' view can help, but is not entirely satisfactory
because it undervalues long-term benefit and survival. However, the
graphical presentation of D.C.F. analyses does give a very useful picture
of 'entrance fees' that are, as we have seen, of increasing importance in
choices of technological opportunity.

High interest rates and rapid inflation still further increase the emphasis on the short term in D.C.F. models. Attention is therefore being switched to developing dynamic models. These assume that the main aim of a technological company in the private sector of a mixed capitalist economy is to survive. Under the present rules of the game this may be interpreted as a requirement to pay a dividend to its shareholders which at least remains constant in real terms, and preferably grows. Corporate behaviour which consistently fails to do this will eventually lead to bankruptcy or takeover. But in the long run a growing dividend payment can be supported only from a growing assets base. In addition, sales and assets growth provides a means of accommodating productivity gains without encountering the social difficulties of managing a sharply decreasing workforce.

A fruitful approach to this type of survival modelling is to use the framework of the annual accounts. In this way the company can explore the relationships between its assets growth, its profitability, and the rate of increase of its gearing (i.e. whether loan capital is rising faster than total assets), and how these might alter if more shares were to be issued. Once a suitable growth and gearing policy is chosen, then the required D.C.F. rate of return for typical capital projects can be calculated. Proper treatment of inflation is vital, particularly if the inflation rate is itself changing, but the overall result is that a company can see the scope it has for investing in ongoing businesses rather than in single projects, knowing that it is not over-reaching itself financially.

This overall survival approach can on occasions reverse investment decisions based purely on D.C.F. analysis, to the long-term benefit of the company concerned. But again, as for D.C.F. techniques purely non-financial considerations such as social responsibility cannot be included easily.

In the public sector, models are distressingly difficult, because of the absence of unifying criteria. Sometimes central governments spend money in pursuit of re-election; sometimes they are seeking to relieve hardship or injustice; sometimes to repel putative invaders; and for much of the time, like their local counterparts, they are just muddled. Often, money is not the best scale to use: important Treasury models have sought to correlate growth with unemployment figures on an empirical basis that of course shifts from year to year.[10]

In discussing the technologist's constructive response to capital shortage (p. 97), we have noted the way in which capital-intensive innovation is followed by imaginative improvement work that makes

113

the resultant mature technology much less capital-intensive; the investment per unit output in the earlier days of electricity, plastics, fibres or airlines was much higher (in real terms) than the average investment per unit output of these mature industries now. But the process of this improvement has been somewhat wasteful; some plants and processes were introduced that either failed or proved incapable of economic development. Further, the course of replacement of one technology by another, which is not the instantaneous affair often visualized, but rather a process taking perhaps twenty years or more, must be one that can be predicted and planned. Because invention and discovery themselves have a strong random element and depend on brilliant perception and an element of accident, attempts to subject them to planning often drive away inventive people or dull their efforts. This sometimes leads to a general reluctance to plan and model the overall process of innovation and improvement. Yet this is illogical: the initiation of human progress may depend on George Bernard Shaw's Unreasonable Man[11] (who expects society to adapt itself to him), but its *prosecution* depends on Shaw's Reasonable Man (who expects to adapt himself to society). Accordingly, provided that invention itself (which is nothing like as expensive as development and innovation) is left reasonably free, there is everything to be said for the modelling and planning of improvement, substitution, and their relationships with innovation and invention. Such models can make considerable use of the Boston Consulting Group's 'experience curves' (see p. 48) and of empirical equations for technological substitution, now being tried out.[12] If successful, they will yet again provide new linkages between technologists and the people they seek to serve, and will further demonstrate that it is possible to avoid the waste of activity that results when new things are made before detailed consideration is given to the precise way in which they will be used and of the effects of their use.

Underlying all the modelling of investment, there is a constraint which we explored in Chapter 3: economic growth and prosperity is still the only acceptable means for significantly increasing social equality. If economic growth is linked to consumption of resources and hardware, and is seen mainly in terms of more cars, hardware, and clothes, it cannot be indefinitely and freely continued, however one models the capital flows that would be needed. If it is linked with better health, an improved environment, greater personal expression in the arts, and improved social experience, it can do just that. Growth models along these lines can be very constructive indeed, but the mechanisms which could divert postponed consumption into the right technologies still need careful design.

4. Conclusions

The postponement of consumption that underpins technology has always depended on confidence. Since the beginning of the first industrial revolution, confidence has been subject to cyclic fluctuation in a way that is roughly understood (though very difficult to control). Now, however, confidence in investment, and enthusiasm for new technology, seem to be in longer-term decline. This, coupled with fierce demand for continuing increase in material benefits (a process fuelled by continuing expectation of technological expansion and by absence of recent memories of economic hardship), has in Western countries produced rapid inflation that reinforces the capital shortage. Authoritarian top-down societies have greater economic stability, but at the expense of innovative vigour.

Technologists have skills that help to respond to capital shortage by ingenuity and selectivity, but additionally now need to understand and respond to the social changes that are occurring. Modelling helps, but the modellers need to pay increasing attention to human behaviour, new possible patterns of organization, and very rapid changes in attitude. The models must therefore be replaced often.

Human behaviour and organization

1. The state of the world

(a) *General*

In tackling resource problems, the technologist faces a need to turn his skills in new directions; for capital and finance problems, he faces public doubts about his ability to continue producing benefits that are sufficient to justify saving. In both areas, his knottiest problems are with the behaviour of people who are much more educated than formerly, but are as yet insufficiently educated to recognize the greater need for social co-operation. Technological growth has brought its own kind of constraint because of greater numbers of people, which result in crowding and aggression; greater power concentrated in the hands of small groups of people on whom large numbers depend; and greater freedom from fear about the consequences of not respecting the *status quo*. The key point about all these factors is that they do not arise because of what technology is or does, but because of the pattern of people's response to it.

Cultures of multiplying micro-organisms stop multiplying more usually because of the accumulation of toxins and growth inhibitors than because of lack of nutrients.[1] Technology seems to behave similarly: the acquisitiveness harnessed by capitalism to finance innovation and investment becomes more destructive in later phases because of the self-confidence and chauvinism generated by its material success. Yet technologists have been slow to face some of the human problems they have created; their success has depended on and has reinforced a specialist system that has licensed them to regard people as inanimate thermodynamic assemblies of customers, clients, operators, and educators, always anxious to help in order to produce and enjoy economic growth. Control of the negative general effects of selfishness has been left to others not very knowledgeable about the technological changes impending: to policemen, judges, doctors, and administrators. Nowhere is modelling and understanding more needed, and nowhere is it so primitive. But a little progress is being made, and the most important task for this book may be to help awaken technologists to this lacuna in their perceptions.

One of the reasons for the dissocation of technology and social skills is, of course, the absorbing claims of time and energy made by the demand to make things or machines effective, cheap, and acceptable. Everyone cannot do everything. But another is a special scientific heresy, explicitly stated by Lord Kelvin: 'When you can measure what you are speaking about, and express it in numbers, you know something about it, but when you cannot measure it, when you cannot express it in numbers, your knowledge is of a meagre and an unsatisfactory kind. It may be the beginning of knowledge, but you have scarcely in your thoughts advanced to the stage of science.' Countless technologists and scientists have framed this statement and hung it above their desks, like a religious text. Some social problems can be assisted by the use and processing of numbers; studies of demography, as we shall see, are a case in point. But far more are concerned with the behaviour and influence of individuals affected by local events.

Quantification has its place in society, of course. At one time, the importance of 'great men' was overemphasized by historians, who ignored population numbers and economic forces and simply told stories about kings, barons, and rebels. Then there were instances of overemphasis on mass imperatives arising from mass dialectics, as in early Marxism—Leninism. Now there seems to be a reconciliation which recognizes the historical importance of individual behaviour, which, however, can be effective only if it either fits with current group perceptions or can be made to generate them.

It is people, not numbers, that have ideas and emotions. It is probably true that the genesis of a branching chain of persuasion requires a group of critical size at the start. The efforts of a single leader, however energetic or frenetic, may die away if he treats his followers as machines, whereas the efforts of a small number of disciples may be enough to light the flame of a new religion. The processes whereby people can be set moving in parallel or complementary activity are subtly different from those for similarly organizing molecules. Each individual can be listened to and understood, so that those arguments are brought to bear that will most effectively deal with his particular pattern of hesitancy and reservation, and those new possibilities pointed out to him that will seem particularly important. He can be relied upon to remember something of what was said while the prophet is speaking to others, although the better and more economically each case has been presented, the better will be the memory and recall. And when, finally, all the disciples are assembled, a group message can be devised that will accentuate the most positive drives and avoid the most tangled remaining negatives. One disciple will hear it differently from another, but both will go forth and testify.

There is nothing wickedly manipulative about such behaviour: it rests on respect for individuals and the widely different genetic structures and remembered experiences that they carry. But it is singularly difficult to generalize in the form of numbers and equations in order to teach technologists. There are indeed generalizations that can clarify personal and group interactions, but they usually rest on qualitative rather than quantitative models. However, since no less a person than Lord Kelvin said qualitative models were unsatisfactory, the technologist may well say that he will not bother with human relationships and interactions until they can be quantified, and that he will thus leave the (as yet unsatisfactory) social 'sciences' alone until they are like physics. Such technologists would do well to remember that the steam engine preceded the complete statement of the laws of thermodynamics on which its operations depend.

Quantitative models of behaviour may make better progress than at present seems likely, but there can be no doubt that technologists can no longer afford the expense of the Kelvin heresy. Unless behaviour can be better understood, even if only qualitatively, and internal inconsistencies can be clearly presented, then populous technological societies must either become authoritarian or destroy themselves. Further, the authoritarian solution (which simplifies behaviour by restricting it to command and compliance, with no complaint or feedback) merely seems to substitute long-term instability for short-term instability; dogmas, dictators, and hegemonies do not seem to last for more than a few centuries and often for less than that. Power really does seem to corrupt.

(b) *Specialism and the size of the disruptive or effective group*

Since 1945, a vast change has occurred in the minimum group size for social effective or disruptive behaviour.[2] Up to that time the practice of war involved the assembly of larger numbers, better weapons, or a more powerful economy than the opposition could muster. It required national unity and skilled leadership to make those large numbers effective, but the combination of threat to homeland and presence of a reasonably resolute and fair king would usually do the trick (hence, of course, it is more difficult to hold together an invading than a defending army). The same has been true in peaceful activities: the economic power of United States agencies has, despite trust-busting, concentrated the world's most powerful computer and aircraft firms in the U.S.A. Only by a much more tolerant attitude to monopoly have European countries managed to match the size of American firms in some other fields, to preserve some innovative competitiveness. Even so, only the incipient unity of the E.E.C. has really kept a balance.

Technology has always been able to place power temporarily in the hands of a small innovative group, in peace and in war, but despite the daunting experience of being attacked by early tanks, machine guns, or bombers, defensive procedures were worked out that made large numbers once more necessary. A major success for technology combined with small numbers was the radar-fighter system devised for the defence of Britain in 1938–9; indeed this could have been assembled by quite a small nation. (No such 'small nation' bomber system was devised: cities are simply too big for significant effects to be achieved with chemical explosives plus small numbers of aircraft).

Since 1945, however, the potential impact of small groups has increased on all fronts. A small nation can now assemble and deliver a nuclear missile, provided that it has unity of purpose and high technology; Israel, with less than four million people, will undoubtedly attain this. It is difficult to say whether a large nation would be able to stop the arrival of at least one or two thermonuclear weapons from such a source. With ingenuity on the part of the attacker, there must be a fair chance of disaster big enough to make counterattack, however massive, by no means a satisfactory riposte.

The more likely and relevant examples for the technologist, however, are in peacetime. Small groups of specialists or of determined militants can bring very large industrial assemblies to a halt. The farming out of automobile component manufacture to specialist firms has emphasized this trend, so that it has now become an aim of technological mangements to escape from such specialist strangleholds, and equally an aim of militant unionists to create and preserve monopoly specialisms. Strikes by clerical workers who collect bills could clearly create crippling cash crises, and strikes by garbage collectors, sewage-plant operators, hospital workers, and doctors further emphasize the interdependence of society. The introduction of computers and instrumentation has created improved control systems, so that plants can operate better and road systems carry more traffic because of computer-linked signals. This creates secondary concentrations of power among the few who are needed to run and maintain such systems; serious congestion, affecting very large numbers of people, can accompany breakdown. The increasing pressure that can be created by a successful action highlights the ultimate need to recognize interdependence.

At present there is far more emotional response to the abuse of specialist power than there is consideration of the options available to run a society created by specialist innovation. This emotion response is particularly evident where groups such as dockers or miners have turned the tables on employment systems that were formerly somewhat tyrannical. There has been very little consideration of satisfaction with work, or of motivating factors. The basic assumption has been that bargaining between acquisitive people will create a system in which those who can improve the system most will acquire most. The entrepreneur who invents and introduces an assembly line, in place of batch workshops, will be able to pay his employees more, and thereby compensate for the loss of the craftsman's delight in making, or identifiably helping to make, a recognizable whole. Even having paid out more, the employer himself will make more, so that all have won (so the argument dangerously runs).

In other cases, specialists have been rewarded by social esteem, so that they have developed a sense of vocation that has transcended monetary reward. Nurses and hospital workers have been underpaid and (within limits) liked it; until recently such jobs represented a low-cost route for climbing from working-class anonymity to middle-class individuality and esteem, as well as providing outlets for altruistic devotion to community service. For a long time, senior civil servants and service officers received part of their reward even more often explicitly in non-monetary terms, with orders and honours. The increasing similarity between employment in the public and private sectors, and pay equalization, has made this progressively more odd.

Since neither the free market nor the social-esteem model now succeeds in preventing specialists from making use of (or abusing) the monopoly power with which a kindly system has provided them, technologists must change their ways. Either they must cease to depend on specialism and operate at 'workshop' level with generally educated people, none of whom has special bargaining power or the power to obtain it, or they must actively pursue methods for conducting specialist activities in ways such that internal satisfactions are high and the perceptions of interdependence are strong enough to ensure consensus in the pattern of rewards. Either is a tall order, but qualitative models can help, as we shall see.

What cannot be successful is for the technologist to take water, wash his hands, and say that all of this is the affair of others. It is far easier to design good systems into social practice at the outset, than to try to clear up appalling conflict that need never have happened.

Repeatedly, human considerations will in future cause procedures to be recommended that are sub-optimal in the narrow, inanimate technological sense. No technologist will readily accept such versions of his work if he truly regards them as sub-optimal. He must therefore be actively and keenly involved in human design and assessments and, in this sense, respect individual perceptions and aspirations as something to be built in.

(c) *Population growth, concentration, and aggression*

The technologist's development of specialism has created rapid population growth by the improvement of medicine, public health, nutrition, and hygiene; and the technologist has himself come to depend on the urban aggregation which ensued. But he has rarely asked the question whether some such aggregation might be more self-destructive than others, or whether technology, having passed through a phase of urban dependence, could now become rural again, or at least micro-urban.[3] He has, however, accepted in a very short space of time—about twenty years—that an arrest of such population growth is probably the most urgent immediate need, and he has made excellent progress in decoupling childbirth from the fulfilment of the sex drive. Again, however, his technical success is frustrated by human factors. Too often what seems best for the individual in fact harms society overall. Just as the British coal-miner seems to gain from militant pay claims (thereby rocking the economy on which he depends), the Indian peasant seems to provide best for maintenance in old age, which he now more often attains, by a family of strong sons (thereby provoking intolerable population growth).

It is possible that the Chinese have already begun to come to terms with the effects of technology on population and its aggregation by intermingling urban and rural life and interchanging many people frequently between the two. In this they have been assisted by a less individualistic social tradition, perhaps reinforced by genetic factors. On the face of it, dispersal of people and their time into agriculture is wasteful, because agriculture can if required by conducted with very few people. However, it may be possible to boost food production very considerably by employing more people than at present away from towns. If so, and if this pattern of employment is satisfactory to the individual, then some escape from the animal aggression of large assemblies may also lead to the solution of other problems. But it remains to be whether such developments can ever be reconciled with Western individualism.

Aggression is also a reality that the technologist has not yet faced up to. Aldous Huxley's *Brave New World* invention of the pacifying and harmless drug 'Soma' is not yet a reality; all current tranquillizers and similar drugs that act on the central nervous system are to some extent habit-forming, less effective under prolonged use, and unsatisfactory for general social use in various other ways. Tobacco smoking and alcohol consumption have their corresponding defects, although both seem acceptable for marginal use. But narcosis seems to imply an authoritarian state, with the apparent consequent certainty of long term corruption and degradation, and Aldous Huxley, associated it with a society designed round a deliberately created class structure and consciousness.

(d) *Progress with the pattern of education*

Two key social impacts of technology, namely population growth and a massive increase in the proportion of people living without major discomfort into old age, have led naturally to a third. This too shows signs of turning into a constraint. It is the great increase in the proportion of people being educated,[4] and the proportion of the life-span regarded as normally devoted to education. This change has been partly due to the availability of time and resources for teaching and learning (i.e. the creation of opportunity for education), and partly to the smaller amount of manual effort, and the larger amount of highly skilled effort, required for technological systems (i.e. the creation of direct need for education). Most technologists have not yet seen the great need for better human systems, but the younger educated members of Western societies certainly have; in consequence they have turned from specialist science and engineering to (newly specialized) social studies, or to arts courses. There is a great ferment in progress.

The results are variable. In Japan, most young people express dissatisfaction during education, and then, as if by a switch phenomenon, turn in employment to the loyal support of a highly organized techno-logical society. As students they are critical of a Japan heading for impossible resource demands and general creation of economic disorder elsewhere, but as workers they conform. In other Western countries, those industries providing basic essentials and affluent goods are tending to attract only the more orthodox students. Those who are most clearly aware of the human problems tend to take jobs concerned with immedi-ate relief of poverty and hardship, sometimes relying for the long term on a shaky political alliance with militants concerned more with raising the wages of their own group than with general problems. The able scientists tend to stick to their subjects. In consequence, specialist education, originally brought into being partly for effective learning and progress in technology, now largely fails to motivate anyone who might understand coming technical events to think about possible solutions for their social consequences.

There is a great need for a mutual alliance and understanding between the young technologists and the young sociologists, for the educational system will take at least as long as the industrial and govern-ment systems to diagnose the troubles. Group learning may have a bigger part to play, and individual or group teaching a smaller one, for the ability to build on one's own individuality in combination with those of others is one of the crucial requirement.

2. Present preferences

Any brief account of the human constraints on technology must in-evitably be unsatisfactory, for they are so widespread. The subject covers the whole span of the mismatched expectation and performance that underlie inflation and the shortage of capital, bad resource and energy policy, inadequate attention to the environment, and the absence of an adequate consensus for effective lawmaking and enforce-ment. The encouraging sign in the present state of the world is the growing but still small number of people who are becoming convinced that maximizing intake of material possessions is not only an unneces-sary and unrewarding exercise, but is increasingly an impossible one to take much further.

This said, there seems to be no reason to reverse the primary human desire that we identified earlier: the reduction of drudgery associated with everyday affairs. However, just as this needs to be pursued with increasing concern about energy consumption, it also needs better handling in human terms. There is no inherent reason why every family should pursue the objective of a house extensively filled with machinery whose use tends to create loneliness in the home[5] and heavy demands on water, roads, and other services. Acceptable long-term household models may indeed consume quite substantial amounts of energy (but no more than can be conservatively generated) and offer considerable access to equipment—but in a pattern that creates communities, while respecting privacy.

Civilizing transport systems will be an important need if the present preference for personal vehicle is to be reconciled with available energy supplies and road space. Ten years ago, the Buchanan Report *Traffic in Towns*[6] showed that unrestricted use of private cars in very large cities was not possible, even with ruthless and expensive clearance for urban road systems. At present, the urban scene is made unnecessarily unpleasant by the congestion and competition involved in moving cars about and parking them. Education and technology for more and better telephones, more precise appreciation of the benefits of self-restraint in urban car use, and an alleviation of the sense of self-importance that usually underlies haste would all help to make the desire for cars still less acceptable.

3. Synoptic 'top-down' modelling for human systems

A proper treatment of this subject would have to deal with religions, political systems, and indeed all centrally designed and run institutions from schools and universities to armies, navies, and tennis clubs. It appears that precise synoptic models like Christianity or communism have to be enforced and kept pure by authority in order to unite action usefully; but we have already expressed dislike for the authoritarian approach. The function of synoptic (top-down) models therefore becomes that of uniting synthetic (bottom-up) models, or of providing guidelines for their construction and use. Several examples may be quoted.

The first is that of the British Commonwealth. It arose not from a grandiose concept of world domination, but from a finding that the separate government of Scotland, England, and Wales seemed inconvenient. Wales was conquered by England; so was Scotland, but the Union of Crowns took place a full three centuries later, when it was convenient to solve the English succession problem that had festered for two centuries by appointing the Scottish King to the English throne. International rivalries within the U.K. were seen as a serious danger to stable government and prosperous trade, and the political and economic benefits of unity (in particular the operation of a common exchequer and defence system) seemed to outweigh the attendant loss of face and local autonomy. After the turbulent Stuart monarchy from 1603 to 1688, the Scottish uprisings backed by the old church stood no real chance of success; for in spite of successes in northern rural areas, the English townspeople saw the advantage of the new unity quite clearly enough to turn out and defeat the rebels.

It was, therefore, quite natural to extend this pragmatic and beneficial model of progressively expanding political union to overseas conquests. The resulting bilateral trading patterns between Britain and her colonies enormously benefited technology in the U.K. by ensuring plentiful flows of cheap raw materials in one direction and cheap manufactured goods in the other. Had anyone modelled the new textile mills, railways, foundries, and engineering workshops, he could have drawn the sub-models together by the Commonwealth concept in such a way as to show that the Doctrine of Comparative Advantage truly worked. The resultant hierarchy of models had politically elected figures at the head, and achieved compatibility by steady delegated work.

The establishment and operation of an imperial model of this kind can be criticized as outright and cynical exploitation by the colonial power, praised as the means whereby education and technological development were disseminated, or observed with neutrality as a phase in geopolitics. What is undeniable is that it kept a series of internally consistent and varied sub-models in operation for longer than would have been the case in its absence. Commonwealth preference assured Britain of cheap primary products, and other territories of a particular market and of cheaper manufactured goods from that source; at one time it also created a series of military alliances that, for better or worse, prevented the substitution of other alignments. Latterly the commercial disadvantages of these arrangements to former colonial territories were considerable. They helped to stop the primary producers combining in order to use their growing economic strength in bargaining for manufactured imports from advanced nations. They also tended to encourage specialist and cash crop developments, and to prevent the emergence of cohesive and comprehensive national power. For their declared and intended purpose, however, they were effective and, equally important, when obsolete they could be discarded without a global upheaval.

A more general but equally pragmatic synoptic human organization model is that of the free-market economy. This need have no military enforcement, although its defence is not infrequently invoked in support of national military enterprises, such as those of the U.S.A. in south-east Asia recently, and formerly those of Britain. It operates in limited conflict with imperial models, since it visualizes world free trade as the ideal, and it favours any producer who can acquire a monopoly or local part-monopoly. For this reason, it requires various legal constraints on monopoly. These constraints, which have considerably affected the technologist, are selective; in the case of new inventions they are temporarily withheld by the patent law to encourage invention. It is thus a model with considerable complexity and subtlety. Again, it favours the *status quo* and is therefore inevitably a target for radical reformers. Whether desirable or not, it is a tried and tested model that has under-pinned the industrial revolutions of the nineteenth and twentieth centuries.

The contrary model of Marxism has provided greater social stability in conjunction with a different pattern of personal freedom and restraint. It has so far proved to be less favourable to major and varied technological innovation for the consumer and, as mentioned, has so far operated only with authoritarian backing. The model of the 'continuing revolution' in China is an extremely interesting dynamic concept that could, in principle, lead to some of the withering away of the state that is a feature of theoretical Marxism not so far translated into practice.

One area of quantitative modelling in human organization that is now becoming important is demography. Quite small changes in rates of input or output can greatly change population characteristics. If birth-rates change with respect to death-rates, or recruitment into a company with respect to resignation and retirement, then age-distribution, promotion prospects, and opportunities for new ideas are all affected. If there are larger numbers of older people who may oppose or delay new ideas and innovations, there may be an increase in social instability. Models of such systems enable policy options to be highlighted. But, as always with top-down models, conclusions as to *what* can be done usually say too little about *how* it can be done. This calls for more detailed modelling at a different level, and it is to this bottom-up modelling that we now turn.

4. Synthetic 'bottom-up' modelling

Machiavelli provides a masterly collection of linkages between upward and downward modelling. He observes certain lessons from his own experience, such as that mercenaries are useless if they lose and fleece their employer if they win. He further proceeds to show that citizen armies, while ideal, preserve their loyalty only to an hereditary ruler, and that a newcomer who is successful in war has trouble in the ensuing peace. He then builds an intriguing synthetic model from which he analyses instances and gives advice.

We now have a much richer collection of synthetic models than could be constructed in the Renaissance, though most of these dealing with human behaviour and organization are still necessarily qualitative. Beginning at the individual level, there are behaviour modes that can help in the building of constructive relationships and avoidance of conflict, and improve comprehension[7,8] (see Appendix). Next, there are conflict models that elucidate competition and can help in controlling it so as to make it constructive; the same models can, of course, also be used simply to win, which is why many of them were first built. Then the generalization of conflict models can sometimes show the limits of competition and the areas of need for regulation and cooperation, together with the concomitant possibility of conspiracy. Regulatory models then lead inevitably to the models of taxation and economics discussed in the preceding chapter, and to the legal models suggested in Chapter 8. Between the downward and upward models lie the models of international agencies and businesses that run across government, and form an important part of the teaching of Galbraith.[9]

(b) *Models of competition and conflict*

The Boston Consulting Group model of competition based on learning curves (see Chapter 4) is an interesting case of benefit to individuals that can be destructive if carried too far. When one is seeking the most effective deployment of a new technology, the pursuit of variants that will increase demand is helpful in reducing costs and pressing ahead. When the technology is mature, competitive action and reaction are less clearly advantageous to society. The pursuit of a high market share that will bring lower unit costs may tell a newcomer to price so as to fill his new, big plant; simultaneously, it will tell an established producer to price so as to keep the newcomer out.[10] If both have considerable reserves of money, the ensuing price war will deprive the industry of the margins needed to maintain a programme of continuing improvement and (if needed) growth. If carried to its logical conclusion, it may end in total withdrawal by all producers but one, who thereafter is no longer stimulated by competition and can be restrained (but cannot be easily stimulated) only by law.

The opposite of a socially disastrous price war is agreement between producers to maintain high prices. Fortunately, there is a range of competitive behaviour in between these extremes that usually combines inventive stimulus with reasonable security of resources available for improvement and growth. Monopoly legislation rules out much price collusion, and commonsense discussions of a general kind, often helped by government agencies, often prevent price wars. In this middle ground, modelling of competitive situations can be assisted by games theory. Rapoport[11] and others classify conflicts as fights (which aim at destruction or elimination of the opponent), games (which aim at evidently outwitting him), and debates (which aim at persuading him). The social aim is to guide conflict along this sequence, away from fights towards debates. Since simple games can be arithmetically formalized, by setting-up a so-called 'pay-off matrix', it is possible to make rough predictions of the way in which the game will settle down if both participants play safe or if both play adventurously.

The difficulty about such numerical procedures is their inevitable dependence on the Kelvin paradigm. This is their social limitation. It is possible to deal fully only with zero-sum games (in which gains of one player are exactly balanced by losses of the other), and the play-safe (minimax) criterion. But there are nevertheless two ways in which attempts to use games theory can help. First, the presentation of situations in the form of pay-off matrices greatly helps to ensure that the nature of the competitive situation has been carefully thought out and defined. It is probably correct to say that a businessman who has written out pay-off matrices wherever possible will have a clearer view than his competitors who have not done the same. Repeatedly, it helps to construct matrices of this kind in relation to personal plans, for it highlights the best options and the greatest probabilities, and encourages the elimination of unpromising and unlikely possibilities.

Secondly, the definition of a zero-sum game can help us to see how it might shift to become a negative-sum game (in which both parties lose) or a positive-sum game (in which both win). Some of the positive-sum games, such as an agreement to rob the bank, are forbidden, but others can turn attention to options that are globally attractive. For example, competition in a mature area may have reached the point where products and processes are sophisticated: natural commercial development at this stage whereby there is concentration by some producers on the manufacture of particular intermediates or components, with free trading in these items, need not inhibit constructive competition in the overall assembly and selling processes, but can help to strengthen drives for improvement or economy.

If new technology, and saving, continue to be less relentlessly pursued, it may well be that competition will and should diminish. In such a case, society will move anyway from games towards debates, and to socially approved decisions as to who makes and does what. But some competitive stimulus should still remain. The technologist and his numbers do have something to offer, but only if he uses them in social models of the kind we have indicated.

(c) *Larger models of communication and organization*

The table of contents of the *The Prince* is, par excellence, a list of organizational models, and the chapters themselves are commendable succinct runs of these models (Table 7.1). There are, however, no rules for the construction of such models except the general one of keeping them simple and avoiding the hopeless pursuit of meaningless quantification. It is dangerous to use numbers whose uncertainty or inexactitude means that apparently precise conclusions involve paying attention to differences well inside the experimental error.

One area of organizational modelling of special importance is that of telecommunications. If people living far apart arrange a meeting (with great effort and expenditure of energy) perhaps once a year in order to collaborate, it is practically certain that most or all of their meeting will be occupied by irritated misunderstanding and conflict, together with its partial resolution. Of true co-operation there will probably be little. Correspondence is not an effective means for clearing up such misunderstanding, for it is not sufficiently interactive, and the meaning perceived by the reader is often not the meaning intended by the writer. The disciplined use of telecommunications for more frequent group discussions can do much to make the meetings, when they occur, much more constructive. But not enough is done to design and use terminals—electrowriters that permit diagrams to be drawn during conversation, or rapid document transmitters, which can enrich conversation. Much emphasis is conventionally placed on video-images which require band widths equivalent to fifty or more scarce telephone channels.

Telecommunicative behaviour has the potential to change society much more than most technologists realize. In the past sixty years the real cost of a long-distance call within the U.K. has fallen ten-fold, and that of a transatlantic call forty-fold; the real cost of a letter has risen over the same period. Although it is an expensive investment, telephone equipment lasts much longer than, say, a motor-car, and uses far less energy. A much greater use of wall telephones without handsets could, for instance, enable young mothers in separate kitchens to chat while minding the children or doing other jobs; once the appropriate social conventions and skills existed, such behaviour would reduce the demand for second cars without increasing the problems of isolation.

Behind the failure to demand and use such facilities lies a failure to model the reasons why telephones are found a nuisance, and to correct the nuisance. Telephone respondents either interpose a telephone dragon (or secretary) between themselves and callers, or they do not. If they do, callers are daunted and inhibited by the barrier. If they do not, a call made during a discussion with someone in the room thrusts the

Table 7.1. Contents of Machiavelli's *The Prince*

I	How many kinds of principality there are and the ways in which they are acquired
II	Hereditary principalities
III	Composite principalities
IV	Why the kingdom of Darius conquered by Alexander did not rebel against his successors after his death
V	How cities or principalities which lived under their own laws should be administered after being conquered
VI	New principalities acquired by one's own arms and prowess
VII	New principalities acquired with the help of fortune and foreign arms
VIII	Those who come to power by crime
IX	The constitutional principality
X	How the power of every principality should be measured
XI	Ecclesiastical principalities
XII	Military organization and mercenary troops
XIII	Auxiliary, composite, and native troops
XIV	How a prince should organize his militia
XV	The things for which men, and especially princes, are praised or blamed
XVI	Generosity and parsimony
XVII	Cruelty and compassion; and whether it is better to be loved than feared, or the reverse
XVIII	How princes should honour their word
XIX	The need to avoid contempt and hatred
XX	Whether fortresses and many of the other present-day expedients to which princes have recourse are useful or not
XXI	How a prince must act to win honour
XXII	A prince's personal staff
XXIII	How flatterers must be shunned
XXIV	Why the Italian princes have lost their states
XXV	How far human affairs are governed by fortune, and how fortune can be opposed
XXVI	Exhortation to liberate Italy from the barbarians

other person in the room rudely aside, angering him and embarrassing the telephone respondent. Consequently there is no possibility of a relaxed conversation. The startlingly simple solution is for all callers, always, to begin all calls with the question, 'Are you busy? If so, when can I ring again?' With such a habit, the telephone would cease to be an embarrassing and hostile device and would become a friendly essential— as it is to an articulate young mother, housebound by young children.

5. Conclusions

Technologists have been inhibited in the vitally necessary consideration of the emotions and reactions of the people they are seeking to serve and help. They have failed to distinguish and tolerate the less precise and more responsive techniques that they must use in human relationships. People are different from things. Yet technologists have encouraged specialization and have thereby placed power in the hands of small groups, greatly complicating society. They therefore have a duty as well as a need to use appropriate and approximate versions of their powerful modelling techniques to study and begin to solve problems of human behaviour and organization.

8

The social control of the environment

1. The state of the world

Greater industrial activity and increased urban size and concentration have made existence noisier, more violent, and more subject to effects from effluent and contaminant waste materials. Simultaneously, greater affluence has liberated resources wherewith such problems, and others concerned with the hazards of the natural environment, can be tackled. Technologists, as we have seen, must be vitally concerned with the preservation of an acceptable and, if possible, an improving quality of life in this difficult and challenging situation. Conventional ideas about priorities in amenity seem unreliable and ephemeral. The device that is acceptable when owned or used by a few can be deafening or asphyxiating when used by the many. Contamination of air, water, or products may already be kept to commendably low levels by dedicated work, but if it is true, as suggested, that a significant proportion of the increase in cancer is due to environmental factors, then there can be no complacency.[1]

Environmental problems cover a wide spectrum, at one end of which they are of natural origin, and at the other of human origin. Their control involves science, engineering, and, most important of all, legislation. Yet the impact of these remedies varies greatly across the spectrum, so that the environmental improver has a difficult task in his choice of weapons. At one end of the spectrum of natural environmental threats, he is unarmed. He can do virtually nothing about bursts of radiation from sunspots, or about the ultimate calamity of the expansion of the sun to become a red giant. He can do nothing as yet, to control the weather, and little enough to forecast it in order to take precautions. His powers begin to be significant in the control of atmospheric quality, although if he is concerned about the upper atmosphere, including the ozone layer that protects against some harmful radiation, he may need world wide collaboration or legislation, because the effects of human activity in a particular place are rapidly disseminated to the rest of the globe. Cooperation and consensus are key elements in any policy seeking to restrict air pollution by dust or volatile materials, or river pollution by liquid effluents.

Detection, the tracing of cause and effect, and the enforcement of remedies often involve difficult science, expensive engineering, and complex argument. Then, in the area of noise and violence in cities, the law is the main corrective instrument. No environmental improver will achieve much unless he is skilled both in the technological control of things and in the sociological understanding or legislative control of people.

This is one of several subjects where there is now serious concern about erosion of freedom. At one extreme, those so concerned are manufacturers who take a narrow view of their social responsibilities, and believe or urge that environmental concern is exaggerated and perhaps economically irresponsible. At the other, there are anarchists who believe that society and its technology are sinister and insidious enemies of the balance of nature and of personal liberty and who oppose large-scale systematic organization, preferring small scale autonomy. In between lies a generally and reasonably worried populace, critical of the bureaucracy, taxation, and waste that large societies seem to generate, and yet uneasily aware that a small number of the ruthlessly dirty or violent can poison society or reduce amenity significantly. One thing is certain; more effective and extensive laws, and bigger and more difficult systems, are needed to sustain the ten thousand million people who are likely to exist early in the twenty-first century than would have been needed to sustain the very much smaller number who were alive three centuries earlier. But the fact is that freedom has fluctuated throughout history. The technology that creates the greater environmental and social constraints also liberates energies and intellects that then cry out for personal achievement, fulfilment, and liberty. A population that must till the soil from dawn to dusk to avoid starvation has little time or concerted effort available to challenge or overthrow its feudal lords. Consequently, we in our generation should not feel particularly badly treated. There was no golden age in the past, and there is probably no golden age to come in the future. Our efforts, however, can probably help strike a good and improving balance between opportunities and problems, both greater than ever before.

The changing response to novel risks can be demonstrated by the safety records in old and new industries. A public outcry results if one nuclear power-station worker dies of leukemia, but that is small comfort to the widows of the tens and hundreds of trawlermen or agricultural workers who are still killed each year. Most employees in modern technological industries are already safer at work than at home or on the road, yet far more money is spent on factory safety campaigns than on domestic ones. Similarly, new drugs or agrochemicals based on good research can, when properly used, yield benefits that far outweigh the dangers and discomforts of not applying them, but they must undergo trials that their more dubious predecessors of fifty years ago were never subjected to.

Life is never free from risk, and the technologist must see that he is not unreasonably handicapped by environmental or legal constraints trying to make it so; this in turn implies that the must understand what society is really trying to achieve in its sometimes capricious and conservative reactions. Complete freedom to disturb the environment with noise, effluents, and all forms of cheap or harmful junk and to ride roughshod over out-of-date laws and regulations while doing so is clearly not to be desired. But the technologist is in danger of being seen (wrongly) in just that light, and curbed over-enthusiastically as a result. It is good publicity material to show one photograph of a stagnant foam-covered river, full of old motor-tyres and other rubbish, alongside another of a pleasant wooded valley with a clear and sparkling stream running through it. The boring, dangerous and poverty-stricken nature of life before the Industrial Revolution is conveniently forgotten, and the technologist is blamed solely as the bringer of pollution in all its forms. Indeed, the concentration on pollution is worrying, since it is to the exclusion of the many positive contributions that the technologist has made to the environment. In this chapter we are therefore forced to explore pollution in more detail than a truly balanced treatment of the environment would merit, simply because that is the sin we are charged with most often, and the area where new laws will impinge most forcefully.

(a) *Pollution: general*

Every living object pollutes its environment by rejecting material that its own metabolism cannot reprocess. Indeed, pollution is almost a defining feature of life itself: as already observed, micro-organisms usually cease to multiply exponentially well before nutrients are exhausted, simply because of accumulated pollution. If populations remain small, and the surrounding fluid flows (air, water) are adequate, then pollution is dispersed. This is a process that can continue for a very long time if the total capacity of the dispersant is large, as it usually is; for the biomass of the earth is confined to a narrow surface layer by comparison with the great volume of the air and seas. If this 'natural' pollution does accumulate, another species often exists whose meat is the polluter's poison; the plant and animal kingdoms, which exchange oxygen and carbon dioxide, are cases in point. There has, of course, to be a free energy input somewhere in all these cycles, for life systems are not perpetual-motion machines; photosynthesis represents the prime energy uptake that is crucial to all terrestrial life, directly or indirectly, and which therefore effectively fuels natural pollution in all its forms.

Man is the first species that deliberately processes and sweetens its ordure and domestic waste. He came to do so because of the economic advantages of community living, and of the experimental finding that his undispersed pollution is dangerous to himself, since it harbours and encourages pathogens and their carriers. Man is probably also the only species to have an aesthetic objection to his own waste and to remove it because it looks or smells bad. In addition he landscapes and uses artificial odours; he takes steps to shut out noise pollution; he sprays water to lay the dust, and operates fountains to sound and look pleasant, and to cool courtyards. If tidiness and aesthetics are to be praised, man is praiseworthy.[2]

Unfortunately his other priorities create increases in domestic and industrial pollution that outstrip his available will and energy to tidy up. For a typical city, rubbish is likely to increase in proportion to built-up area (if technology stands still), or as an even higher power of its diameter (if technology advances). The rate of operation of the dispersants (air, rivers etc.) tends to increase only as the first power of the city diameter, i.e. as the square root of its area. But artificial rubbish processing which minimizes the need for dispersal (e.g. tipping, incinerating) tends to annoy the immediate neighbours, as well as being expensive.

The main problem of pollution, then, is not that it is something man is particularly bad at dealing with: it is that he has increased unit sizes too much and too quickly for his existing cleaning-up procedures. The purpose of developing the argument along these lines is *not* to urge that pollution problems are so exaggerated that, if left alone, they will go away. It is to point out that one of the most important anti-pollution steps at present would be to limit the size of cities, industrial plants, and numbers of vehicles at some specified level and then work on the pollution problems at that level.

Pollution is one of the diseconomies of scale, and a society whose units are all being scaled up will have pollution problems that are intensifying more severely than the scale-up rate. The first need is to help existing agencies and improve existing methods. To speak of pollution as it it were a recently discovered problem would seem to impede the remedies by causing confusion; where is the sense in urging the British Leyland research department to invent the wheel? To be sure, there are criminally negligent polluters, some of whom should be jailed; but more will be achieved by spurring and encouraging existing efforts and their supporters than by guillotining factory managers and owners in the market square. The most effective course is to devise models and procedures that demonstrate how tidiness pays by channelling the benefits from such work to the tidier.

There is little point in tabulating or cataloguing pollutants, for everything is a pollutant in sufficient quantity or in the wrong place: water in the form of a tidal wave, air in the form of a tornado, or even a spilt lorry-load of gold or powdered streptomycin, if it blocks the road. Some substances, such as lead, plutonium, mercury, some polypeptides and some proteins, can pollute at very low concentrations; these require special attention. But over-emphasis on spectacular and specific current problems should not occupy all our available attention, and it seems sensible to begin by considering the methods of attack on pollution that have already enjoyed great success.

(b) *Attack by limitation of scale*

Pollution is by no means the only diseconomy of scale. Unit distribution costs from a single source to a wider range of consumers rise as the radius of the market area increases; large work-forces tend to build up tensions, disputes, misunderstanding, frustration and inefficiency more often than small ones; a large unit may be a more conspicuous target for political criticism because of suspicion (perhaps quite unjustified) of monopolistic and unfair behaviour, or undue use of local influence. If there *is* monopolistic or strong-arm misbehaviour this is probably also a diseconomy in the long run, because although buccaneering may bring short-term advantages, society often catches up, puts on severe and penal constraints, and leaves them on for a long time.

Economies of scale usually tail off when units become very big. On the other hand, we have seen above that the diseconomies often rise more steeply as size becomes really large. Fig. 8.1 shows this illustratively. If, even in rough-and-ready qualitative terms, it can be seen from the study of cities, factories, or ships that there is a minimum in the sum of the effects of scale, with a fairly sharp increase thereafter, it makes sense to try to arrest growth in size before the point where diseconomies become serious. Otherwise, very damaging individual responses can occur. Cities of more than one or two million people certainly fall into this class: congestion, violence, and the high costs of urban management are beginning to drive out those people who can get jobs elsewhere, and these are usually the ablest and most energetic residents. Factories that are too big are prone to strikes; both big cities and big factories have more severe pollution problems. The water closet, although perfectly acceptable in towns of scores of thousands, set bigger problems in the collection and disposal of sewage from very big cities. There are methods for tackling big units and their problems, but there is now enough evidence to show that some of the problems can and should be tackled before the big unit is even built.

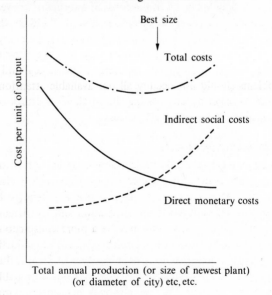

Fig. 8.1. Economies of scale.

(c) *Attack by the development of scavengers*

Pollution, especially if it is of biological origin, can often be dealt with by deliberately providing systems in which scavengers flourish. In wet

137

countries, micro-organisms are found that deal with most organic pollutants, and they can be adapted so as to make them more active and effective. The design of hold-up vessels, in which scavengers can operate, is of course the basis of all sewage treatment, and has been used for factory effluents ranging from thiocyanate from gasworks to phenols from chemical plants. In dry countries, insects can sometimes be found and developed for use: Australian dung-beetles accumulate and roll-up animal excrement conveniently. More sophistication is being used in such work, to which microbiologists and entomologists are making substantial contributions.

(d) *Attack by using pollution as a resource*

Unsolved pollution problems can be regarded as the generators of a negative resource. Just as positive resources can often be upgraded, so negative resources, thought of in these terms, can sometimes be turned positive. This then provides purely economic incentives and assistance for pollution control. Some industries have a strong tradition in this field: the chemical industry, for example, tackled the old Leblanc process (which made gaseous hydrochloric acid and evil-smelling crude calcium sulphide as by-products of sodium carbonate) by devising two quite new processes for economically converting the unwanted by-products into valuable chemicals: sulphide into sulphur or sulphuric acid: hydrochloric acid into bleach.[3] Waste dirty sulphuric acid has been used to make ammonium sulphate fertilizers. The metal industries are full of instances of the beneficial use of slag (e.g. for fertilizers) and tailings (for recovering other metals). The micro-organisms that grow as part of sewage treatment can be harvested as protein for animal feed, provided that the sewage is not mixed with metallic or other poisons; paper-mill effluent is already being utilized in this way.

Recycling of paper, old motor tyres, and waste plastics provide further examples. The sum total of achievements is very considerable. A crucial requirement sometimes is organization to segregate different kinds of waste, for sorting of unnecessarily mixed rubbish can be very expensive. It is important also to remember that the traditional laws of supply and demand apply just as much to pollutants as to any other commodity. If too much waste paper is collected for recycling then the price will fall until consumption can be increased.

(e) *Attack by process improvement or product change*

All technological developments that are designed to give higher-yielding processes, or to avoid the wastage of materials as unwanted by-products, contribute to pollution control. This covers a very large range, such as diecasting instead of making shapes by machinery, or greater selectivity in chemical catalysts. Even the pork butcher's traditional boast of using everything but the squeak, although praiseworthy, can in principle be simplified by devising a process for making bacon-flavoured microbial protein and carbon dioxide from sewage and recycling all the process liquor except for a small purge used to fertilize the factory rosebeds. Product replacement can have even more drastic effects. The use of a small part of television receiving and processing space to build up static images could readily provide enough room to carry all the pages of a newspaper without the use of newsprint.[4] As we have already seen, the replacement of personal travel by the educated and sophisticated use of telephones, with electrowriter terminals to enable diagrams to be drawn and seen, could eliminate great *tranches* of resource consumption and pollution.

(f) *Strategic attack: positive and negative attitudes*

Taxation methods and incentives can play a powerful part in saving resources and reducing pollution. In the last resort, prohibition and prevention all have their place in strategies for pollution control. But much more effective and extensive pollution control has been achieved by treating it constructively, as an opportunity for the increase of benefit, rather than as an opportunity for launching political attacks. Thus it may well seem theoretically sensible to tax equipment that merely burns electricity, rather than using it as a source of motive power; this would presumably cut back the growth in the demand thus reducing thermal pollution of rivers by extra power stations. One could argue for a tax of £100/kW, say, so that any necessary extra generating capacity could be financed largely from the sale of electric fires. But such a campaign would certainly not be helped by slogans such as 'make polluting pensioners pay', and opposition of a highly emotional kind would certainly (and rightly) defeat the whole programme. Conversely one could seek the reduction of aircraft noise and fuel waste constructively by advertising that 747s cause less disturbance and need fewer airports and less fuel per million passenger miles than do Concordes or first-generation small jets. This will have more effect than describing Concorde passengers as public enemies, when their offence is in fact a normal human desire for distinction, from which most people suffer to a greater or lesser degree.

2. The law: general

Technologists are profoundly affected by and involved with the law, and increasingly so as their activities penetrate more deeply into everyday existence. Many past laws clearly have some bearing on technology[5] but, since they were drafted under very different circumstances, require interpretation that leaves room for argument and doubt. New laws are equally often difficult to draft in such a way that they give adequate protection to society and yet do not strangle technology completely. The man with a red flag who was required to precede early motor-cars ensured safety but prevented a development that society seems to have wanted. The setting of toxicological limits and the clearance of new drugs repeat this problem many times.

This is not the place to list the laws that may constrain technologists, but they affect all the constraints dealt with in this book: ownership of and rights to resources and energy; probity and good management in relation to the custody of capital that has been borrowed; relations between different members of the community and the rights of individuals and groups; and protection of the environment. Protection of technological property is partly dependent on the rights and duties imposed by the laws of patents and monopolies. International and maritime law play a part too, and any technologist responsible for work in several territories may have to consider each separately.

The crucial decision the technologist must take, again and again, is whether he regards the law as an ass or as reasonable. For the law changes continually, and represents a shifting consensus; there are seemingly no longer any general perceptions that legal principles are absolute. With changing consensus, for example, suicide and abortion have passed in the West from the forbidden to the tolerated region; and in the absence of adequate support, laws about trade union behaviour, although passed by a British government with a parliamentary majority, have proved unenforceable.[6]

The technologist has a number of choices. He can observe the law; ignore it and rely on his ability to escape detection; argue inapplicability; seek repeal; try to assemble a consensus of militant objectors that will make the law unenforceable; or take his operations to countries that suit him better. All are done at various times. He can also seek to be involved in the drafting and formulation of legislation in order to improve his compatibility with the law. If he can help the lawyers to devise workable rather than unworkable statutes, he can often have quite a considerable say in their shape. As an example, laws are sometimes passed demanding zero emission of some pollutant. Now zero emission is clearly impossible, since all emissions can, for argument's sake, be held to be certain to contain the occasional molecule of any substance legislated against. In practice, 'zero' must be interpreted as 'undetectable', for enforcement demands unequivocal detection. But levels of detection can change drastically because of quite external analytical developments, and are entirely unrelated structurally to levels of danger. Consequently limits related to detectability represent an unpredictable threat both to operation and to consensus support; it is much better to state specific levels that permit rational and honest response, as in the recent control work on vinyl chloride toxicity.

The most unsatisfactory situations develop when technologists, for one reason or another, take the line that the law is an ass and seek to frustrate or evade it by 'fine print' arguments. One particular problem arises over the attitude and capabilities of enforcement or regulatory agencies relating to toxicology, factory safety, or finance. It is these, rather than criminal or civil courts as such, which represent the closest legal links with the technologist. On the one hand, it may become clear to the members of such an agency that positive clearance of a new development is always dangerous to their reputation or careers, for they may then be proved wrong and therefore castigated as rash; extreme caution, however, and even deliberate obstruction are more rarely damaging to them, for the technologists consequently frustrated often have very little political power. In such a situation technologists may tend to work to the letter rather than the spirit of the law and fight rearguard actions against all new protective measures. The resulting technical disputes are often inconclusive and are usually expensive: settlement has to be arbitrary and is consequently often unjust.

On the other hand, a protective agency may become too responsive to the alleged difficulties of those to be controlled. This can be reinforced by a situation in which the agency is undermanned, so that it can always claim, in respect of any alleged failure of duty, to have been busy elsewhere on matters of higher priority. This leads to public indignation and allegation of collusion, and in due course to harsher and less acceptable regulations than would have resulted from a less comfortable but more defensible control. The enforcement agency thus sails between the Scylla of neurotic negativeness and the Charybdis of sloppy permissiveness,[7] while the technologist must navigate between the cliff of harsh restraint, demanding the seeking of unintended ways through, and whirlpool of public allegation or suspicion of corrupt collusion with the controllers.

Little modelling of the law and the interactions of its microscopic provisions has yet been done. In a way, this is odd, for although English law explicitly eschews the top-down model, Roman law, the basis for the Scottish, French, and many other systems used such a model in a primitive way. It may be that the top-down study of interactions between law, technology, taxation, and behaviour will be possible when the English Law Commission, now ten years old, has completed more of its global reconciliations and rationalizations.

All these observations indicate that it is very difficult to translate a public wish that 'there ought to be a law against it' into an enforced reality that a law exists whose operation strikes a good and effective balance between safe and frustrating stagnation and dangerous but often beneficial change. Again and again, it proves to be best when the technologist and the law-maker join forces with the consumer in setting and achieving an accepted compromise. There are three parties to to system, not just one or two, and none of them must be ignored, given totally free rein, or deceived.

3. Present preferences

The present separation of technological and non-technological education plays a large part in generating tension around environmental and legal issues. Society tends to want a better environment than it is prepared to pay for, and technologists tend to want more freedom than they are prepared to earn by work and explanation. If the two forces lack a mutually understood language (as they often do), then expensive dispute, avoidable disasters, and lost opportunities are inevitable. The crusading environmentalist wishing to swing public opinion has only to find one buccaneer out of a hundred technologists earnestly co-operating with the authorities. Furthermore, since the disasters (such as thalidomide) produce results that rightly excite compassion, while the lost opportunities (such as a candidate oral contraceptive for males turned down on some inconclusive animal tests) are never seen, the dice are loaded in favour of restriction. Even this, however, is not all: when restrictions prove to be dearer than was realized, they tend to be dropped completely. This may happen with the United States legislation about automobile exhausts, which was drawn up with only limited consideration of sustained feasibility, or of increased calls on fuel and hence on scarce capital for more refineries. In the resultant disillusion, what would have been a practicable, enforceable, and acceptable intermediate policy is lost altogether.

Such arguments are not put forward in defence of myopic technological optimism, complacency, or a view that controlling agencies should get off the back of technology. Environmental improvement must be increasingly demanded, as must better laws; both must be paid for, not only directly, but with much better understanding and communication.

4. Synoptic 'top-down' modelling

The areas of environment and law benefit notably from time to time by co-operative top-down models that catch the public imagination and unite it with technological possibilities and imperatives. One reason for such success is that there is a strong reinforcing effect between individual success in this area; the effect of ten district clean-air schemes in a city means that each area does much better than it would have done alone. Equally, one demonstration, by refusal to comply with a majority-supported law, that such laws can be defeated, damages the credibility and strength of the whole legal system. Further, it is possible to pick objectives which, if well-timed, can command almost as much support as Christianity did in medieval Europe. Some examples may help.

Clean-air schemes in the U.K. are an area where environmental benefit has arisen apparently by legal means, but where in fact the complete top-down model has included in addition several non-legal elements. The London smogs, particularly in 1952 and 1955, were a prime example of conditions in which the fluid above a vast city was quite unable to cope with the effluent being pumped into it; in another sense they illustrate the well-known effect of post-disaster legislation, in which one single catastrophe catches the politicians' imagination in a way that a recurring nuisance never can. But economic and social factors were as important as legal and enforcement factors in ensuring that the resulting clean air laws were a success. First, for twenty years from the early 1950s onwards fuel-oil and gas prices moved downwards relative to that of coal, partly as a result of poor economic bargaining by the oil-rich countries (prior to the formation of OPEC) and the introduction of a petrochemical and then a natural source for gas. Then too, the general desire for less drudgery played a part, since there were no longer housemaids to lay open fires in every room of every middle-class house before breakfast and to brush away the resulting coal-dust daily.

Another less obvious example comes from the world of the motorist, who since 1938 in the U.K. has by and large complied with laws restricting the unsociable sounding of his horn. The model here is of better roads, safer motor-cars, and chiefly a much greater traffic density, rendering the original message of the horn unnecessary in modern conditions. So long as there were only some hundreds of thousands of motor vehicles on Britain's narrow roads it was both necessary and courteous to sound your horn to say, 'I am here behind you'. Now that the chief preoccupation of the British motorist is with other road-users, rather than with merely propelling and steering his vehicle to his destination, the horn is relegated (or promoted) to occasional use, often as a device for insult or castigation, but rarely as a warning. Thus the fact that it is illegal to sound it after 10.30 p.m. in a built-up area can readily be accepted, except apparently in London's West End. Indeed it is arguable now that the law is superfluous, having spurred the motorist in a direction in which he was already moving, and having ensured the better sleep of those who live close to main roads.

The wartime waste paper campaigns commanded great support, and although there is considerable need for similar operations now, the top-down model is different and acceptance more difficult. Then the populace was waging total economic war in a way that its successors are not; faced with newspapers that were flimsy and small, with slim books poorly printed on semi-transparent brown paper, and even with the disappearance of the local park railings because of resource shortages, the sensible and indeed pressing requirements for retention and recycling were obvious. Now they are by no means so apparent, when each family suffers from a surfeit of newspapers and magazines rich in advertisements, and even fish-and-chip shops use virgin paper to protect their wares instead of old copies of the *Daily Express*. The campaigns of thirty years ago needed no legal backing; yet now the law alone would be unlikely to reproduce similar behaviour, because the parameters in the model have changed.

5. Synthetic 'bottom-up modelling

Successful top-down models that unite diverse parties in complementary and synergistic actions are, not unnaturally, the exception rather than the rule; further, they depend on happy timing, and can rarely be invented on demand. Yet the power of the modelling approach will ensure that models will be built. Such bottom-up models will usually arise when one interested party seeks to understand and strengthen his position in the system. The remainder of this chapter develops some examples of such models.

The areas of resource conservation and pollution control are, as we have seen, very closely linked, for positive attack on particular environmental or waste problems typically increase the stock of useful resources. A scheme for harvesting leaves to recover protein reduces the number of fallen leaves to be swept up, and straw recovery reduces the amount of rural acrid smoke in autumn; similarly, well-designed rubber or plastics recovery can save hydrocarbon production. There is considerable danger, however, that recycling the wrong part of a total system may prove to be wasteful of energy or of labour. The first modelling need, then, is to design production and recovery cycles that totally economize on resources, with no illusory response to slogans about automatic benefit from procedures involving recycling.

Like most economic calculations concerning social activities, these models can be slanted for special sectional pleading. There is likely to be a need for generally trusted impartial agencies, which, however, must be competent and listen to all arguments. It is possible to pillory pesticide manufacturers and down-rate their modelling because of profit motives, while trusting the calculations of bird-protection societies, who are not motivated by profit. In fact, both are sectional interests whose conclusions are in need of independent scrutiny. It should be noted, however, that the manufacturers will probably have more money than the ornithologists, which may make their sums either more thorough or more deviously misleading.

We are here venturing into the comparatively new field of *technology assessment,* which lays great stress on completeness of approach. But even the Office of Technology Assessment which provides an independent service to the U.S. Congress is usually forced, as are similar organizations in the Swedish cabinet office and the State of Victoria (Australia), merely to accept and scrutinize submissions from interested parties. Provided that the right questions are asked of the models this need not be a great handicap.

Recycling models will by no means command automatic acceptance. The simplest method whereby an extraction or production industry, whether capitalist or socialist, can pursue self-interest, esteem, and influence, is by pursuing output growth. Growth not only enhances economic power directly but also offers various routes to the reduction of unit costs with the consequent possibility of yet more sales through lower prices, or more internal financing of more research or investment. Recycling may therefore be seen by such an industry as a threat, since it may erode the growth toward which everyone has been urged in the past. Arguments such as those of the Boston Consulting Group (Chapter 4) urge everyone to race down the learning curves of escalating output.

The conservation of resources and of the environment may, therefore, suffer a head-on collision, of a quite non-political character, with technological tradition.

One encouraging sign comes from the metal industries, where recycling has never been resisted, and where most the work, employment, and profit margin comes from fabrication rather then smelting. This effectively decouples profit and esteem from sheer volume of production. If the grocery trade were to adopt a model that sought to apportion and protect more food with the consumption of less oil, then in the plastics processing industry the amount of added value per ton of plastic (film) could be increased. Very thin, oriented films demand more care in manufacture, and if the industry became more deeply involved with thin film packs rather than bottles or thick packs, then lower proportions of the employment and margin would come from the basic manufacture of resin. The difference here is between making 'things' that have value because of what they do, and 'materials' that have value because of what they consist of.

Progress toward better total systems will, however, be very slow if environmentalists and manufacturers first develop and then argue from the basis of different synthetic models, without reconciliation. Ideally the mutually trusted intermediary discussed above (perhaps a knowledgeable university or research agency financed independently) might unite all those concerned round a model that recommended policies acceptable to everyone. If this cannot be done, external influences may have to be applied. One such would be a differential tax on resins used for packaging otherwise than as thin films. Such remedies, however, are always blunt and are on occasion liable to produce the wrong result.

The sub-models so generated have to be considered in terms of overall distribution. Which is the more economical in oil or capital, a large deep freezer in every home coupled with sales of large quantities of food needing less packing material, less labour in subdivision, and less petrol for transport; or central packaging plus supermarkets and normal refrigerators or larders? This could cause reversals of some conclusions. Then again, models of varying extents of 'eating out' or 'eating at work' would further be needed. There might be no unique solutions, but long-term trends could be influenced with much more understanding, and policy actions would generate fewer unpleasant surprises.

Environmental and legal modelling is greatly needed in urban development, although it is very difficult to make it quantitative, and there have been serious disappointments. However, qualitatively it is possible to see a good many points. For example, the cheapest method at present for constructing an acceptable dwelling is to do the job in a factory; after which the cheapest method of installation is to put the unit on wheels and trundle it on to the site. This, however, limits the height to one storey. Cabin construction and simple assembly from trucked units can give two-storey construction, but anything further begins to require technology and equipment at the site.

The American mobile home industry accounts for more new units annually than does fixed construction, and the standards are far above those of the traditional U.K. holiday caravan. But compactness is still at a premium. Mobile home economics are also, obviously, favoured by the low land values that obtain in large countries with less pressure for agricultural land. However, large assemblies of mobile homes on site are unattractive and easily become squalid. While favoured by the 'second-home' or 'pipeline-laying, and weekly movement' models, they are not favoured by a 'suburban amenity' model. Little is left for the purchaser to do himself, so that the pride that goes into making curtains, fitting shelves and furniture, and other do-it-yourself activities, has no outlet. Satisfaction for urges, creative or otherwise, tends to be driven out of the home, and sometimes there may seem to be a mismatch between a mobile home (which can be moved) and a garden (which cannot) to the detriment of gardening. Consequently, with high land values fixed construction is favoured.

Other models then bear on the choice between medium- and high-density housing; if the latter is chosen, there are still high- or medium-rise choices. Traffic models favour medium-density housing, which creates less severe parking and peak flow problems. Psychological and behavioural models tend to favour medium-rise building of a kind that encourages neighbourhood interest in and protection of the approaches. A sense of ownership is important: violence and vandalism seem to flourish particularly where no one can see the approaches to apartment entrances in high-rise buildings.

The assembly of an appropriate housing policy from all these models is also complicated by the fact that the best compromise may not encourage imposing architecture and may therefore not be rewarded by architectural prizes. A variety of neighbourhood disasters have been generated by conformity with architectural paradigms that have rewarded architects but penalized those who had to live in their buildings. Yet a further constraint to housing is, of course, the formal expression of past model solutions as codified in local building by-laws. In spite of continual revision these cannot always reflect the opportunities presented by new materials or technologies, which are sometimes thereby hampered.

But the main constraint is inevitably human prejudice and lack of imagination. This is something we seem to have overcome better at some periods in the past. It is arguable that one of the most successful housing models, both in its own time and for succeeding generations, is the large Georgian terrace house. One of its main strengths is its flexibility: originally built for families with servants, many examples exist that have since been used successively as flats, offices for a firm of solicitors, surgeries for a group dental practice, and so on. Provided that a mews was also part of the initial model even the catastrophic arrival in our cities of the motor-car has not rendered this solution obsolete. Yet how many private dwellings built today, the confused production of technology, architecture, by-laws and tradition, will still be praised for their beauty and versatility in another 150 years?

Environmental traffic modelling is a well-established and important subject, embodying operations research, queuing theory, noise studies, and many other exercises. Control systems can be designed like those for plants and factories, and again hierarchies are important. The arterial motorway that speeds up inter-urban transport may deliver vehicles at its exit at a rate beyond the capacity of the capillaries leading away from it, and it is so much more expensive to solve the intra-urban problems that they either go unsolved or have to be tackled by limitations on parking that prevent full use of the capacity of the arteries. An important early study of a series of traffic movement hierarchies in towns of increasing size and difficulty was provided by Colin Buchanan's 1963 *Traffic in Towns,* which studied Newbury (population 30 000, so that planning for unrestricted use of cars is possible) Leeds, and London (where planning for unrestricted car use is impossible).

6. Conclusions

Society's comparatively new-found but praiseworthy concern for the environment seems to make it necessary for the law to say 'halt' or 'slow down' to technologists, who often resent these instructions. But the co-operative use of modelling techniques—part synthetic, and part synoptic—seems to be particularly necessary to avoid dangerous compromises in planning houses, towns, roads, rivers and waste management, and to create a common perception that generates co-operation rather than conflict. The law, if resorted to gleefully as a weapon against political enemies, does not work well now that society contains less structural fear. In its application to technology and the environment, effort devoted to increasing the majority who understand and support the law must pay off handsomely in decreased need to invoke it, and in increased effectiveness when its application is essential.

9

The strength of the complete technologist

1. Modelling with audacity

In the preceding seven chapters we have sought to do three things. First, to note that the technologist faces more critical analysis and less enthusiastic acceptance than before, and to point out that this marks the end of a 150-year honeymoon. Secondly, to observe that he already needs to build an array of skills into his work, particularly those of human perception and understanding; for increased speed, size, or ingenuity will be sought or paid for by society only if they bring demonstrable and sustainable overall benefits. Thirdly, to indicate that although the world is no longer his oyster all is not yet lost to the complete technologist. His skills are urgently needed for the transition from non-renewable to renewable resources and energy sources; for the more effective and sparing use of savings and capital; to design the human application of all technology with greater sensitivity; and to help in the creation of better environments and laws for a world population that technological skills will have allowed to grow to well over ten thousand millions.

Most importantly, his methodology in building small models for the operation and understanding of limited areas of work—a plant, a factory, or a traffic system—and then assembling of strategy, should have very broad application in social design and operation. Political theory has in the past largely depended on broad conceptual generalizations which tended to confuse and divide people if not authoritatively interpreted and enforced. This has certainly been true of the medieval Christian and the twentieth-century Marxist states. Overall concepts and models still help, but they need more careful reconciliation with the realities of everyday systems. The technologist's patience and ability to do this are among his most valuable assets and contributions. He is accustomed to the use of algebra, calculus, or qualitative and descriptive models for a wide variety of situations, and he is not restricted to stereotypes such as 'monetarism', 'capitalism', or 'communism', whose characteristics and patterns have to be defined in immutable or historical terms. He is accustomed to building dynamic models that interpret behaviour in time, and he can even build in discontinuities. He is used to the disconcerting effects of error and uncertainty, and to the utility and limitations of methods for handling them. Above all, he has experience of disciplined experiment, of orderly retreat where there are real obstacles, of methods for climbing hills in the dark without a map, and of pragmatic exploration punctuated by the prudent cry, 'Silence, while the leader advances.'

If the technologist reading so far feels reassured as to capabilities and possibilities, all is well. If he is beginning to feel complacent and confident, he should pause, for we were speaking of the *complete* technologist, caparisoned at every point, and few can claim to be thus qualified and experienced. Those who are involved in affairs, and are accustomed to argument and the political control of reality, usually lack the best skills in experiment and modelling; and those who are practised in modelling have usually tested their skills more on things than on people. There are very few nearly complete technologists; indeed, if we put an absolute interpretation on the word 'complete', there are none. Consequently, to use and benefit from these potential capabilities, we have great need to educate ourselves and each other.

Technology has depended on the juxtaposition of imagination, free-ranging curiosity, and inquiry on the one hand, and disciplined implementation of patterned instructions on the other. It has also depended on the social reward of directed acquisitiveness and greed, constrained by social customs that have restricted the greatest rewards to a minority of people. Fairly obviously, there has been steady pressure (and indeed encouragement) to escape from the ranks of the commanded and disciplined minions and, by education or use of social monopoly, to join and stay in the commanding élite. The distinctive privileges of this élite have had to decline as its numbers have grown. Equally, the directed greed whereby a small number have generated or invested in innovation has tended to become more disruptive as it has become more general. A democratic élitist society risks declining toward anarchy by reason of its very success in generating numerous élite groups; society becomes increasingly dependent on each of these groups and each of them seeks uniqueness of personal privilege.

The few nearly complete technologists, then, will need to ally themselves with a similarly small number of nearly complete sociologists, and look with some concern over a sea of chauvinism. The degree of acceptance for their proposals will initially be small, and meanwhile conflicts between Maoism, jingoism, conservatism, neo-Stalinism, and other doctrines will produce economic disequilibria and disasters, possibilities of nuclear war, and guillotines in the market-places. The nearly complete technologist, before he has any chance to use some of his skills, may have to decide whether to roll with the punches, dig himself a hole, grow potatoes, operate as an urban guerilla, get himself elected into a government he will be unable to control, or face the sunrise and the firing squad with protestations of loyalty of the creed that has seemed the closest approximation to the output of his models.

Whatever kind of Utopia anyone may perceive (and technologists are more likely to see a signpost saying 'Utopia 10^6 light years' than a door marked 'Utopia'), it is undeniable that we are beginning from a miserable point; most technologists are in the position of the man asked the way to Glasgow, and replying, 'If I were going to Glasgow, I wouldn't start from here.'

The time of writing this book is one of great economic disequilibrium, distortion in the allocation of real resources, and the pursuance of sectional interests that make the situation worse rather than better. The past models of economic equilibrium predict nothing but disaster. Equilibrium in nature is not, however, always attained. If it were, the oceans would react with the atmosphere to produce a nitric acid solution in pursuit of chemical equilibrium, after which there would doubtless be some interesting botanical consequences; nuclear equilibrium would require all the earth's hdrogren to be transformed into iron.

The technologist familiar with these observations may therefore respond differently from the stockbroker. He will know that the Stock Exchange provided only about 1 per cent of the new industrial capital in the U.K. in 1973, instead of the massive inputs presumed by orthodox capitalism, but he will see this not necessarily as a disaster, but rather as evidence that different new agencies may well be needed to secure the flow of capital into novel meritorious enterprises. He may then seek to model or design such agencies, and to consider methods for making them work well and secure acceptance. He may also find new ways for preventing the disastrous attainment of equilibrium in such a way as to halt the trend away from it.

Machiavelli observed that 'fortune' (by which he meant that part of life not governed by his precepts) may well be the ruler of half our actions, but that she allows the other half, or thereabouts, to be governed by us. He went on to conclude that, fortune varying and men remaining fixed in their ways, they are successful as long as these ways conform to circumstances. It is better, he says, to be impetuous than cautious, for fortune is a woman who must be mastered by force. Like a woman, he adds, she is always a friend to the young who are less cautious, fiercer, and master her with great audacity. Technologists, like Machiavelli's Prince, must also reckon with the factors outside their explicit and systematic control.

If we are correct in saying that technologists have some skills in living with uncertainty and in deploying effort on those parts of a system that can be understood and controlled, while allowing properly for the likely vagaries of the rest, then Machiavelli's pragmatism holds out hope to the young technologist if he is bold. For perhaps he can enlarge the area of control, and he can probably do a great deal better by building new models and trying them out than by gazing longingly back at the days when the old models seemed to apply, and backing reluctantly and fearfully into a future where they almost certainly do not.

Let us take an example of a situation where new models seem to be applicable. Modern society has depended for a long time on the absence of some large yet quite feasible money flows; for example, on the bank depositors not removing all their money at the same time. As we saw in Chapter 6, other more desirable flows are now becoming conspicuous by their absence. British industry cannot not rely on massive borrowing from the Stock Exchange, so it seeks to retain enough of its earnings to do without. It therefore recycles money on the basis of perceptions about social and market behaviour that is longer in term than that of a government (which must seek re-election, and therefore please the electors, within five years), and much longer in term than that of the financial institutions. Industrial recycling is, therefore, the principal form of saving in the U.K. that is patient enough for the technologist to use safely in these inflationary times. It is money that moves according to laws different from those of the currency used in the High Street, which represents consumption rather than investment, and different again from those that control the money collected and redistributed by the government. Some of this public money may be used by the technologist, provided that he remembers that he may be expected to produce results soon enough to influence the next election.

The technologist who models this changed situation will recognize that recycled industrial money is definable and limited, and that proposals for using it in new ways must compete with the extension of existing capability to produce food, clothing, housing, or hardware. There is certainly not enough of it for the old concept that 'industry can always find the money to finance a good invention with a promising market' to remain valid. Consequently, he will get the best hearing if he comes forward with ideas that need less capital, simplify existence, or develop more efficient processes and products while adding more value. Direct methods for turning thermoplastics into garments, or raising the fraction of a vegetable plant that can be utilized, or communicating effectively without travel, all make sense in this way.

2. Selection, sacrifice, and niggardliness

This prescription offered here differs greatly from that presently applicable in the United States, Europe, Russia, or China and its acceptance is, of course, a long way off. But it embodies selection and sacrifice, and we may learn from this in our present situation. The selection is the channelling of invention into cleverly identified relevance without loss of flair, brilliance, or surprise. The sacrifice is the abandonment of permission for a thousand flowers to bloom, so that ten fine ones can emerge.

This lesson is a new one. Scientists and technologists have more often than not raised funds on the basis of a claim that their work is something of which society cannot have too much; so that if treasuries are being restrictive, it is permissible and laudable to outwit and confound them by various devices. One of these has been the fear of competition. Persuade the voters that prestige or prosperity demands that we get in ahead of the Ruritanians (who are forging ahead), and the treasury has been outflanked.

Such policies may well have been necessary and pragmatically justifiable in harsher times when technology was smaller and weaker. Historical growth rates for development were not indefinitely sustainable, however, as Lord Bowden pointed out in the mid-sixties: on some spectacular extrapolations a particular technology or science bade fair to absorb entire national incomes by the end of the century.

Some form of selection is consequently inevitable, and technologists now have to decide whether to join responsibly in social deliberations about priorities and allocation of resources, or to stay aloof and continue to buccaneer. The latter policy has some attraction, for at least it provides the alibi 'Please, it wasn't me' when an overmanned development organization is finally instructed to stop growing or (worse still) to cut its numbers. After decades of growth at several per cent per year, this instruction has cruel and uncomfortable effects.

Just as a hard fate may await the messenger who brings the evil tidings that technological growth is no longer unrestrained, harsh criticism from many of his peers may greet the scientist or technologist who becomes socially and economically responsible. Self-imposed restrictions require more courage than those imposed externally. The pursuit of a reputation for liberality, said Machiavelli, calls for continuing display and, in due course, the absence of resources for essential purposes. 'For these reasons a prince must care little for the reputation of being a miser, if he wishes to avoid robbing his subjects . . . and not to be forced to become rapacious; this niggardliness is one of those vices which enable him to reign. There is nothing which destroys itself so much as liberality, for by using it you lose the power of using it.' The technologist does not need to be as ostentatious as a Renaissance prince, so for him Machiavelli's words are too extreme, but they offer an interesting cartoon.

A timely recognition of the arrival of constraint creates discomfort and hard work on new problems. But this is greatly preferable to the cruelty ultimately and inevitably required by way of correction, if there is somnambulistic, ocynical, or misguided overshoot. Conversely, of course, premature restriction and consequent loss of opportunity (perhaps to do something vital, such as to arrive at a new and effective method for population control) is also culpable. The technologist, if he chooses to abandon buccaneering, may not be the best or the only person capable of taking decisions about his resource allocations, but he is certainly the best person to guess at the probability of overshoot or the sacrifices of various levels of constraint. The time has arrived, we believe, when he is at fault if he does not seek widen his responsibilities, to acquire a reputation for being miserly when this seems necessary, and to be accepted as a key co-operating participant in the full range of social decisions, to all of which his skills are relevant.

3. Reinforcement of reform in education

It is important at this stage to ask how technologists might be taught to become complete. Mere exhortation is not enough. Teachers of all kinds have been deluged with too much advice, and have been caused much fruitless work by the imposition of ill-thought-out reform. They have a good record for adaptation and innovation, but change in the nature of their output can arrive only slowly. Attempts at major reform in educational systems are liable to be ineffective and irritating if made in pursuit of only short- or medium-term objectives. Consequently, we now seek only to reinforce those educational measures that have already been initiated, for there are quite enough of these to serve current purpose.

Table 9.1 shows, in outline, some of these activities. Essentially, there are three points of attack; new opportunities in initial education that seek to integrate technological studies with a knowledge of the human systems needed for their effective choice and use; smaller-scale but still important remedial courses aimed at building in additional skills at this initial point of contact; and continuing post-experience education, to enable people to acquire new capabilities after a time in the real world to help them judge what is needed. Equally, there are three areas of endeavour: reinforcement of basic skills, development of specialist knowledge, and assembly of new skills for interesting and socially purposeful studies.

Table 9.1. Areas of current educational development

Areas of endeavour → / Points of attack ↓	Basic education (grammar, rhetoric, logic)	Specialist Studies	Assembly of skills
Initial (secondary and tertiary) education	General studies courses: debating essay-writing	Most of orthodox secondary and tertiary education	Business schools (undergraduate and postgraduate)
Remedial steps in initial education	Study skills units (e.g. A.N.U., Canberra) Remedial secondary teaching	Special courses (e.g. for research students)	Short courses for scientists on their skills in 'affairs' (S.R.C.)
Continuing (post-experience) education	Very little	Sabbatical leave; major refresher courses (e.g. medical)	Business schools (post-experience)

The area calling for least comment is that of specialist studies, for here the existing educational system has been accustomed to pursue excellence for all who can participate effectively. For some little time, rapid specialist advance in most fields has created concern about obsolescence in graduate skills relatively early in specialist carrers. Post-experience 'refreshers', together with assiduous reading, do much to put matters right, and the problem is diluted by the frequent move of specialists into more general jobs (calling more for experience and general soundness of analytical and human skills). If this book tends to concentrate on the importance of building other capabilities round specialism this is not to deny the vital part played by reductionism and studies in depth: technology will perish without them, and social adaptability will suffer severely.

Basic education, once a requirement given considerable emphasis, has suffered from the pressure of specialist advance and the sheer volume of knowledge on all the rest of the timetable. Too many technologists, although well equipped in their own special field, lack abilities in the selective and critical assembly of relevant data, its reduction to a useful pattern, and articulate and effective communication of ideas and conclusions in order to persuade people to do things or pay attention. At one time, proved ability in the 'trivium' of grammar, rhetoric, and logic was a prerequisite for entry into the medieval 'quadrivium' of specialist subjects: music, astronomy, arithmetic, and geometry. Now, however, the very usage of the word 'trivial' to mean 'unimportant' signals danger. There are many specialists who have learnt much of their material by rote (and are therefore deficient in knowledge of its grammar), who have taken arguments for granted (and not probed their logic), and who are poor at explaining what has happened and persuading people to help with the next steps (showing deficiency in rhetoric). If specialist education *really* demanded so much time that little or none was left for acquiring the basic skills, then there would be a dilemma. Since it almost certainly includes a good deal of multiple exemplification of reliably distilled priniples, it is hard to accept that time cannot be spared for the trivium.

What happens at present is that good teachers know all this, and deal with the matter tactically within specialist subjects. A chemist can be taught grammar—both that of his language and that of his subject—by writing essays which are then criticized for the quality and clarity of the connections as well as for the factual accuracy of the content. Research students can be called on to present work at colloquia, and actively encouraged to find out if people understood and enjoyed the occasion. High standards of logic can continually be required. Teachers who devotedly do all this will rightly say that they have subsumed trivial studies within their contribution to the quadrivium; those who do not, or cannot, will often pretend that they too do the job. But there is generally little encouragement or recognition for such dedication when it comes to teacher's pay and promotion, and most would agree that the overall result leaves much to be desired.

We would therefore urge that nothing should be said or done that will discourage those teachers who *do* educate their pupils to be literate and articulate, and everything should be done—including more explicit recognition and rewards—to encourage and help others to do likewise. It is a tedious business to correct such deficiencies in someone over eighteen, for there will often be an embarrassed reluctance to admit that anything is wrong. The setting of essays is not enough by itself, for they may be written as uninspired and uncritical catalogues of accurately reported work, with a need for much wise counsel before anything readable and useful emerges. But there is an addition that can help in this and other cases: the use of group learning. In employment situations, project work requires co-operation and understanding between a variety of specialists, and the presentation of the final choice must be done with clarity and persuasiveness. With such an obvious need, explanation gains steadily in clarity, and a member of the group may emerge as 'trainer' in this respect; sometimes, the project leader can be chosen to play this part. In primary schools groups of pupils often learn in this way, but we suggest there are notable benefits to be gained from the continuation of group project work in secondary and even tertiary education. Superficially, it does not seem to be as efficient as class teaching or group tuition; in fact, good teachers find that, in the right proportions, it is. Where there is an element of the remedial it can increase acceptability, and one might even form quite an effective club of 'inarticulates anonymous' to build on this principle.

All three of the trivial subjects need wider development for modern society. Rhetoric at one time needed to cover little more than conversation, the dialectic, the courtroom presentation, and the market-place address. Now it must also encompass radio, television, the telephone (which is quite different from personal conversation), advertising, the numerous forms of the printed word, and campaigns of mixed method. Better study would improve the effectiveness and honesty of the use of these techniques, and much the same applies to the extension of grammar through systems studies and logic through computer software.

Our third area, namely the assembly of specific skills which already assume literacy, numeracy, and articulacy on the part of their practitioners, is now receiving a good deal of attention. It is usually necessary to have some end-result in mind, for there is no limit to the combinations that are interesting, and human capabilities are not such as to make it reasonable for every able person to assemble everything. 'Interdisciplinary studies', like anything new, can become a bandwaggon or even a refuge for the lazy (following the cynical comment that physical chemists discuss chemistry when with physicists, and physics when with chemists, and women when with both). But new developments in astronomy, for instance, based on great military improvements in the generation and detection of electromagnetic radiation, have required electronics, computers, radio, engineering, and new branches of mathematics, as well as the traditional skills of optics. Later, rocketry, guidance, control, and telecommunications needed to be added. The practitioners of such skills may well have different languages, jargon, theories and even culture patterns and political allegiances. The leader of a cosmology or radioastronomy group will not uncommonly wish to run his own 'reconciling' postgraduate course, to synthesize his new culture. Such initiatives are often exciting and invigorating; more so, perhaps, than the staid pursuit of well-established subjects with their orthodoxies, their hierarchies and sometimes their schoolmen. Molecular biology, immunology, and materials science are further examples that have enjoyed success and gathered able young people. Sometimes an existing subject is reformed and rejuvenated by a new disciplinary slant, as geology has been by the study of continental drift and plate tectonics.

The teachers and students of orthodox subjects—including those whose excellence is essential to the new assemblies—tend to be somewhat resentful of the upstarts. Because of the concentration on new objectives that has brought together the new initiatives, the outputs are often comprehensible and attractive to society, which therefore votes funds more generously. Many chemists resent the view that their study is now a vitally important 'enabling subject', like mathematics, and hence less likely than cosmology or biochemistry to produce great and disturbing surprises. This is like telling an ageing ballerina that she really ought to become a choreographer, or a *prima donna* that she ought to teach singing. It is probable that Ninette de Valois has influenced ballet more than Fonteyn, and that Lehmann did more for opera by teaching Grace Bumbry and others than she did by singing the Marschallin; but most fifteen-year-olds have their eye more on the curtain calls and the deluges of flowers than on the verdicts of the scholars and historians.

The assemblers of skills, then, are often important and influential innovators; but they are perhaps resented at the same time. This is especially true of the new generation of assemblers of technological and human skills. We have seen that most of the technologist's constraints arise in the last analysis from quite understandable human reactions to the new situations and possibilities that he has created. With technological economics, liberal studies in science, systems engineering, and science policy studies launched and away, can purposeful sociology (*not* manipulation but understanding) be allowed to lag far behind?

Disciplinary boundaries, however annoying, are necessary if teaching is to be effective. Consequently, although there can be excursions into interdisciplinary thinking at school and university, there are limits set by the need for firm foundations. The Oxford Philosophy, Politics, and Economics course (and, of course, the great cathedral of Literae Humaniores itself) is an example of a once-new interdisciplinary assembly now blessed as a core study. Physical chemistry and biochemistry have similarly settled down. It is probably best to begin interdisciplinary attacks fairly late in the educational process—perhaps with emphasis on post-experience studies, as in the business schools.

One area of success is the short course, perhaps a week or so in length, to acquaint those approaching the end of their formal education with the breadth and the range of tasks to which they could contribute. Such courses can make use of group discussions, role-playing exercises (like the war games used for so long in military training, and explicitly recommended by Machiavelli to his prince), and questioning of those prominent in affairs—either as representatives (including trade union officials) or as ministers and directors. Another successful area is the interdisciplinary research degree, making a study (which, it is hoped, will be useful) of situations such as the contrast between the success of a particular industry in some countries and its failure in others. Both these approaches receive active support from the Research Councils in the U.K.

In the end, however, the key to educational development is collaboration and understanding between the older and more experienced (who wish to warn of pitfalls and of personal suffering) and the young and eager (who will not suffer avoidable constraint). Very often, younger people are more inclined to respond to past failure by root-and-branch reform rather than by prudent correction. Machiavelli, in his fifties, recommended the impetuousness of the young. Perhaps his example is worth following today.

4. Listening and response

This book is the result of collaboration, spiced by occasional disagreements, between members of two generations, whose dates of birth span a quarter of a century. We have learnt, and continue to learn from one another, in the course of our jobs. We wrote the book to clear our minds, and to establish some discipline, but have been daunted by the range of subjects into which it has taken us and about which we have had to think and learn. This range guarantees that there will be errors of fact and of judgment here and there. Why, therefore, have we been so bold as to publish?

The scientific and technological literature, it can be argued, is too magisterial and insufficiently experimental and inquiring. Many feel that it is a disgrace to be told of an avoidable error that has gone into print. Life is too short to check everything and also argue and discuss broad topics constructively with others: the magisterial approach has the effect of reinforcing an ever-narrowing specialism. We have published this book, therefore, in order to provide a large anvil on which, we hope, people can hammer. If there are too many demonstrable errors, we may have attempted to write too broadly, but the risk seems worth taking. We shall listen most carefully to the hammering; the anvil will have served well even if, at the end, some young and arrogant Siegfried cleaves it with a blow of the sword he has forged on it.

Thus, this book has been written to stimulate discussion, and it would be disappointing if it were received with polite applause or in stunned silence. Our hope is that our readers may respond with messages that will improve the technologist's skills in recognizing and dealing with his problems. We will use these ourselves and, if we write again, will pass them on to our friends in a new and improved version.

Our eloquent predecessor Machiavelli attracted enemies in his own time, and has been misrepresented ever since, in spite of the care with which he avoided inflammatory statements in his writing. We have been less careful, and no doubt our evident distrust of authoritarian solutions will make new enemies for us. But it is urgent to start communicating now; the modelling methods are there waiting to be used. This opportunity must be seized, for if the technologist of the future is not a social animal he is nothing. On this, the validity and durability of his contract with society depends.

Appendix: The use of formal and informal models: error and simplicity

1. Introduction

In the foregoing pages we have tried to advise technologists about parts of their job that are not covered by their formal education, and then to help fill this gap by drawing attention to the existence and utility of ideographic models.

Special recommendation has been given for simple models that can be quickly understood and used. But first priority has been given to the arguments, and description of models and their use has been given second priority in order to preserve continuity and readability. This has led to a somewhat disorderly admixture of formal and well-recognized models (such as those of discounted cash flow or learning and experience) with informal and special models, devised for particular illustration and so far used only by ourselves. It therefore seems worth discussing the merits of formality and informality in models, indicating in a little more detail how formal models are regularly used, or are beginning to be used, in a structured and well-understood way. It also seems useful to give a little more attention to the question of choice of degree of complexity. There are innumerable problems where very complex and specific modelling is beneficial—indeed, the whole of the computer industry depends on this—and our recommendation of simplicity (reinforced here by reprinting a somewhat irreverent article) needs to be backed by some observations on the place of complexity.

2. Formality and informality

The advantage of informality in model construction is that it offers the possibility of precise design to serve the immediate purpose and fit the system under consideration. This is offset by three disadvantages: first, that there is a big teaching job to do because no one will have seen or used the model before and may therefore take time to understand it; secondly, because use and criticism of the model may require corrections that then diminish its credibility and acceptance; and thirdly, that its very excellence for the immediate purpose may make a model less useful for the other cases to which it could apply but which have been ignored in its design. Consequently, any inventor of a new, informal model can gain considerably by designing it so that many people will try it, find it useful, test it, and thereafter educate others. In this way, some loss of precise immediate fit may be well worth while because of wider utility and less need for subsequent special teaching. Informal modellers can thus be expected to peddle their wares and seek to get them accepted into the dictionary of ideograms in common use. Formalization effects the use of computers and the design of systems in various ways. High-level languages make it easier to address the machine with a problem. Compilers enable programmes to be translated so as to be acceptable for other machines. Set suites of programmes, of general utility, become available for specially important problems such as pipework or flow design. Dictionaries are becoming progressively more needed.

Even so, it is to be hoped that it will never become unfashionable to model for one's own purposes. No more need be said on this topic now, for the pattern of use of informal models varies very greatly and is difficult to generalize.

3. Formal models

We shall examine two examples of formal models. The first is a numerical investment model that was originally designed primarily to choose between different propositions, but is now increasingly used to design portfolios of projects that use available cash without seriously risking insolvency. Additionally, it can be used as a management tool, for indicating what combination of future price trends and cost improvements are necessary to fulfil original expectations and thus generate the cash already expected and earmarked. This is a heavy duty for a piece of relatively new language, for it has never been performed in the past, when borrowing from the equity market was open-ended, and shareholders were risk-takers who (in theory) might be disappointed in their expectations, receive low dividends, and respond by changing the board and chairman. The second is a behaviour model referred to as the 'Johari Window' that can be used to improve the degree of co-operation and understanding between pairs of individuals; it has not been extensively used but has considerable potential. Both have been connected with the same scenario, to show how numbers can be used in the first but are neither needed nor helpful in the second. Thus, the Kelvin paradigm (see p. 117) applies in one case and not in the other.

(a) *Investment modelling: combination of D.C.F. and Boston learning models*

As already stated (p. 47), the simplest method of comparing the superficial merits of alternative investment is to construct the standard cost sheet for the operation, including overheads and depreciation charges so as to write off the plant over a reasonable estimated period of obsolescence; subtract the total cost for a year's manufacture at designed output from the sales income after deducting selling expenses, and then express the sales margin so calculated as a percentage of the capital cost of the plant. In addition to this pre-tax return, one can calculate the post-tax return, and then compare different processes, different plant siting, or even investment in different products, with due allowance for different trading risks. The simplest approach here is to estimate the sensitivity of the return to various misfortunes of roughly comparable likelihood—or indeed, on the other side of the ledger, windfalls that might occur. For example, one might write

Project A (plant in Ruritania)

Post-tax return	18%
Chemical efficiency 5% down	−1% on return
Prices 5% less than assumed	−3%
Capital cost up by 10% (engineering or rate of exchange)	−3%
Chance of nationalization in first five years, with adequate compensation	1 in 10
Chance of expropriation in first five years, without compensation	1 in 50

Project B (bigger plant in Barataria, to meet combined demands of Baratarian and Ruritanian markets)

Post-tax return	23%
Chemical efficiency 5% down	−0·8%
Prices 5% less than assumed	−3%
Capital cost up by 10%	−2·5%
Imposition of 10% import duty by Ruritania	−2%
Three-month strike by gondoliers who move product from Barataria	−10%
Chance of expropriation within five years	1 in 10

Having noted that the estimation of the likelihood of expropriation or strikes was speculative and could well be seriously wrong, one might be inclined to concentrate on Ruritania, particularly if a German competitor announced a competitive Baratarian extension. Additionally, the Ruritanian government, anxious to promote domestic industry, might have tentatively offered import duty, a 30 per cent investment grant and a four-year tax holiday, together with 150 per cent tax relief on R and D done in the country. One might then want to add in

Benefit of 10% import duty in assuring demand to saturate capacity
Benefit of 30% grant
Benefit, progressively increasing, of higher throughout by 'debottlenecking' worked out by extra Ruritanian chemical engineers
Benefit of quick plant construction and start-up by extra Ruritanian chemical engineers
Benefit of tax holiday

None of these benefits can be quantified so as to permit their assessment alongside the other sensitivity factors by the method of the single-year cost sheet, because they are all variable in their effect at different times. Whichever year is chosen for comparison, there will be a different picture, and one therefore needs an integrated scenario covering a period of perhaps ten years, so as to include benefits at the construction, tax holiday, and process improvement stages. This brings in the integrated cash flow model, already used on p. 113 (to illustrate the increasing impact of entry fees). Such cash flow models could also include the effects of nationalization with varying levels of compensation, at various postulated dates, and all tax, grant, and fixed payments. Each run, for a particular scenario, can be displayed as a graph showing the early out-flow and later recovery of cash. The imposed boundary conditions can be of various kinds, but a usual procedure is to pick a time horizon before which complete obsolescence of the plant or the technology is unlikely—again, say, ten years. One then demands complete recovery of the investment at that date, and calculates the constant rate of return on the widely varying residual debt that will achieve this result.

Fig. A.1 shows two Ruritanian and Baratarian scenarios studied in this way. This graphical presentation has an extra utility; it displays the time-dependence of the money 'at risk', and the way in which different unexpected events can alter the demand for credit. It is therefore a valu-able component for studies of overall insolvency risks. This is especially important in times of inflation, for delays between outlay and recovery of cash then have effects quite different from those in a world of stable money.

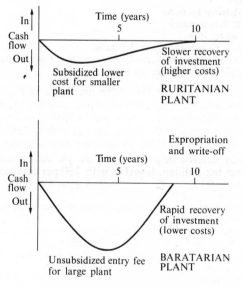

Fig. A.1. An investment model.

The Baratarian scenario involves a bigger initial outlay, partly because the plant is bigger, and partly because there is no investment grant. But the economies of scale then permit the investment to be recovered more quickly, because of lower costs and higher margins. This effect is big enough to enable one to postulate a case in which expropriation in year 9 can be accommodated within the boundary conditions, but with a lower average return. This has been included, but the consequences of periodic strikes of gondoliers have not.

Unless there were important extra strategic reasons for wanting a presence in Barataria, most managements would choose the Ruritanian plant, with its much lower entry fee, higher return when allowance is made for Baratarian hazards, and the possibility of using in other enterprises the capital not committed—perhaps for a giant plant in Riesenheim to serve Nibelheim and Valhalla.

The most recent addition to D.C.F. (discounted cash flow) modelling is the building-in of the expectation of cost reduction according to the Boston Learning Curve (see Fig. 4.1). This would improve the Ruritanian project if the Riesenheim plant were built, for there would then be an expectation of more rapid cost reduction because cumulative tonnage would rise faster, experience would build up, and numerous forms of cost reduction would be financed that would benefit all the plants alike. Such a combination model then provides all those in the business—whether in the U.K., the U.S.A., Japan, Ruritania, or Riesenheim—with targets for cost performance that are ambitious and yet reasonable in the light of the achievement of other, similar industries.

(b) *Behaviour modelling: the Johari Window*

Technologists can feel comfortable with D.C.F. modelling because of its use of numbers and thus its consistency with the Kelvin paradigm (or heresy). But they will miss opportunities if they assess the annual chances of a strike of gondoliers in Barataria at 10 per cent, include this in the sums, and leave it at that. Gondoliers go on strike because of tangible factors such as a desire for better working conditions or the desire of their wives and inamorata for new dresses. Mutual understanding can be actively pursued in many ways so as to remove altogether the likelihood of their striking. Here the modelling has to move away from numbers and Kelvin concepts.

Fig. A.2 shows a qualitative model that can help with behaviour of this kind. The perceptions of the plant management and the gondoliers are plotted on different squares so as to show the degree of understanding

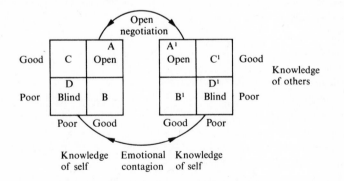

Fig. A.2. A behaviour model.

of each of their own and the others' position. In the upper central squares A and A^1, both understand completely. The managers know that the gondoliers can be satisfied by the means to buy their wives one new dress each for every fiesta, and the gondoliers know that the managers are under reasonable pressure to achieve returns large enough to maintain market share against a competitor. Negotiations can therefore be tough, but the managers may, for example, know that what matters is the Mardi Gras, and that an increase at New Year will give the girls time to make their dresses. The gondoliers know that good results through the next two quarters will assure the position of the Baratarian business. Accordingly, a settlement for a New Year wage increase is found to be just acceptable to everyone, and negotiation is open.

Contrast this with a situation in which the two parties are in squares D and D^1 respectively, with a very poor understanding either of their own or of the other's position. The gondoliers are being nagged by their wives about anything and everything, and especially about rising prices due to inflation. The managers have only recently moved in, do not believe in the cost reductions of which the chemical engineers are talking, and are being badgered about safety statistics, tax arrangements, effluent problems, and other matters. Consequently, everyone's mind is on problems and unreasonableness, and no one is trying to visualize a mutually acceptable solution. Both sides are blind and angry, and the negotiations can fairly be described as 'emotional contagion'. Nothing will be gained by talking; tempers will rise, fists fly, and the likelihood of a strike will increase.

If one manager, or one gondolier, has ever seen or used this diagram (which is known as the Johari Window), he can begin to solve the problem. First, he must learn about his own affairs, and which of them affect the gondoliers. For example, it will be clear that at this stage the gondoliers will have no concern about the directors in London. It is important for one of them to establish that the first priority for most of his friends really is with dresses for Mardi Gras, and for one of the managers to establish that credibility places first priority on good margins for the next two quarters. These two pioneers can then move into squares B and B^1, knowing about their own objectives but not those of their interlocutors. At this point, they can educate one another, and can thus move into squares A and A^1, after which they can work on their colleagues and begin on fruitful negotiations.

Other interactions correspond to disclosure, confidence, or betrayal, and a body of rational doctrine about communication at varying levels of knowledge and effectiveness can be built round the model. With the understanding thus generated it is possible to avoid unnecessary misunderstanding. There may, in the end, be no basis for agreement, but an organized mixture of education and exploration can reduce the incidence of failure through ignorance.

3. Simplicity and complexity

There are no arguments against complexity in models of well-understood physical systems (whether in chemical engineering or cosmology) except those of expense and of the increased difficulty of ensuring that the many assumptions are valid over the whole range of solutions envisaged. Fault-finding in a big systems analysis can be a big job. Probably the most spectacular case of the emergency use of a big model was the exploration of the rescue strategies for the malfunctioning moon vehicle *Apollo 13*; this required all the detail that could the thrown in, and was of course completely justified by the brilliantly successful outcome. Apart from the crucial matter of the lives of three men, American space engineering emerged with a reputation more enhanced than it would have been had the mission succeeded. Since, in retrospect, the moon landing programmes did more for morale and methodology than for knowledge of the solar system, *Apollo 13* may, as an unscripted drama highlighting courage and coolness linked by big computers, be seen as the most interesting part of the whole programme by those who tell the story in the next century.

With this background it may be helpful to reprint a narrative written some five years ago, at the same time as *Apollo 13*.

4. Error and the need for simplicity: Green for Danger

Because of the possibility of shocks, electric wiring is colour-coded. It is therefore deplorable that numbers, which can be the cause of more serious (and unnecessary) shocks, affecting millions of people and not just individuals, are not properly colour-coded. It is true that losses are printed in red and profits in black, but this is not enough. All dangerous forecast profit numbers should be printed in green (with losses still in red), and black should be reserved for the 'hard' numbers that can be really trusted.

Hard numbers are of two kinds. First, there are those that refer to things that have happened and can be measured reasonably precisely— say, to within 95 per cent confidence limits of ±5 per cent. (This may sound far from reassuring to an astronaut or even an airline passenger, who needs something rather better, but it is an accuracy considerably better than that of the 'soft' numbers of which we are now speaking. The second kind of hard number is the technical projection or prediction that depends only on the continued operation of laws that are reliable, applied to numbers that have been reasonably precisely measured. This includes the expected time of splashdown of an *Apollo* capsule, the pattern of the next lunar eclipse, and the tide tables for Pembroke Dock. Even railway timetables contain fairly hard numbers: modern track and electric traction have hardened up the performance considerably. But the monthly U.K. trade balance, with the uncertainty arising from special deals, irregular debt interest, and simple mistakes, is a soft number, even though historical, and should appear in green.

Soft numbers are those that inevitably contain substantial errors, often of a non-quantifiable kind. The most persistent source of softness is the fickleness of human behaviour and choice, enshrined in the measurements of market forecasters, social scientists, and the predictors of dates for plant start-ups. We suggest that all numbers be regarded as soft which contain, in their final form, more than 30 per cent of the error of their sociological component, when this has been compounded. Thus, any prediction of project or business profitability in a highly competitive area consists of soft, green numbers on which business consultants and computer-makers graze and grow fat. The shortfall of profit below the prognostication of the 10-year D.C.F. in the original capital expenditure proposal may be legitimately made the basis for disappointment, recrimination, or discharge (for being unlucky, which in the hard world is a crime), but it is often unreasonable as a basis for astonishment. Predicted profit, which amplifies the uncertainties in forecasts of sales realization, is a very soft number indeed in a dynamic, competitive situation. What, then, is the purpose of helping the colour-printing industry by demanding three-colour treatment of business and other documents? For this reason; that the possession of green fingers for the successful manipulation of green numbers is not just a matter of genetics; education comes in as well. We need horticultural schools in which would-be econometricians can be taught the empirical mysteries of tilth and the country lore that permits some rough-and-ready weather forecasting, so that the crops can more often be gathered in in the nick of time.

Even more important, machinery can now be used, provided that it is the right, robust, coarse, red-painted kind of machinery. Meaningful, rough-and-ready study of broad trends, without too much attention to the actual green numbers, can make coarse-scale economic modelling a very valuable skill, provided that the results are used quickly, before they have become out of date. 'Chairman, a 30 000-tonner is too small. Build her big and fill her up, my old father used to say. The D.C.F.s done on the old 1130 in the back of the haybarn show that a 45 000-tonner would be about right, and you could shut down the little old plants at the bottom of the 15-acre meadow. Another thing; don't let the city-slicker accountants talk you into cutting down on your innovative research, even though the harvest may be bad this year. Some dynamic modelling on this old fertilizer sack shows the drop in 20-year net present worth if you do. Bonnie bairns never come from a house that eats the seed-corn. We could make the old combine last another year, and keep going that way.'

Farmers and market gardeners do not use grey 360/195s that cough and splutter when the air-conditioning goes wrong. The output is no more useful than that from the old 1130, despite the 512K core store. And the rental would be much better spent on soil analyses and dietary supplements. A big computer room looks wrong alongside the privy. But a modern green-fingered modeller can make a great difference to farm profits.

Black-coated, white-collared black-number modellers are, of course, quite right for the rocket base, the warehouse, or the design office. There they can (sometimes) use all of a big core store, and the horny-handed farm modeller cannot do as well. But the black coats and the polished black shoes get muddy down on the farm, and their wearers do not have the advantage of being able to recall wise remarks from farming forebears.

Tractors, bought with sweat and blood, used to rust and rot in Russian fields in the twenties. How many unsuitable computers are now rotting in unsuitable environments or, worse, being used to confuse honest people by the pretence that green numbers are black or that superfluous numbers are potentially useful action numbers? Let the number consumer insist on value for his money and on colour coding that guards him from shocks. Direct access and conversational modes are what is needed down on the farm, rather than remote-terminal, microwave, city-centre procedures.

References

CHAPTER 2

1. Vereker, C. *Eighteenth century optimism.* Liverpool University Press (1967).
2. Shklar, J. N. *Men and citizens: A study of Rousseau's Social Theory.* Cambridge University Press (1969).
3. Wright, E. H. *The meaning of Rousseau.* Oxford University Press (1929).
4. Derry, T. K., and Williams, T. I. *A short history of technology.* Oxford University Press (1960).
5. Schon, D. A. *Technology and change: The new Heraclitus.* Pergamon Press (1967).
6. Kelly, A. *The Relationship between science and technology.* British Association, 137th Meeting, Guildford 1975.
7. Ashby Sir Eric *Royal Commission on Environmental Pollution. Third Report: Pollution in some British estuaries and coastal waters.* Cmnd. 5054. H.M.S.O. (1972).
8. Ehrlich, P. R. and Enhrlich, A. H. *Population, resources, environment.* W. H. Freeman and Co., 1970.
9. Davies, D. S. Discontinuities in chemistry and chemical technology. The Fifth Royal Society Technology Lecture. *Proc. R. Soc. A* **330**, 149–72 (1972).
10. Office of Technology Assessment: *Annual Report to Congress.* U.S. Government Printing Office, Washington D.C. (1974).
11. Meadows, D. H. *et al. The limits to growth.* Earth Island Ltd. (1972).
12. Kahn, H. and Wiener, A. J. *The year 2000.* Collier-Macmillan Ltd. (1967).
13. Fulton, Lord *The Civil Service:* Vol. 1. *Report of the Committee 1966–68.* H.M.S.O. (1968).
14. Rothschild, Lord. *A framework for Government research and development (The organisation and management of Government research and development).* Cmnd. 2814 H.M.S.O. (1971).
15. Jenkins, G. M. and Youle, P. V. *Systems engineering: A unifying approach in industry and society.* C. A. Watts and Co, London (1971).

References

16. Rapoport, A. *Fights, games and debates*. University of Michigan Press (1960).
17. Shaw, G. B. *The Perfect Wagnerite*. Constable (1898).
18. Trevelyan, G. M. *British history in the nineteenth century* (1782–1901). Longmans (1922).
19. Dahrendorf, R. Reith lecture, The new liberty, expansion to improvement. *The Listener* 14 Nov. 1974, p. 622.
20. Williams, J. D. *The compleat strategyst*. McGraw-Hill (1966).
21. Rapoport, A. *Fights, games and debates*, p. 242. University of Michigan Press (1960).
22. Jenkins, G. M. and Youle, P. V. *Systems Engineering: A Unifying Approach in Industry and Society*, p. 35. C. A. Watts and Co., London (1971).
23. Armytage, W. H. G. *Yesterday's tomorrows: A historical survey of future societies*. University of Toronto Press (1968).
24. *Njal's Saga*. Penguin Books (1970).
25. Butler, S. *Erewhon*. (1872). Penguin Books (1970).
26. Verne, J. *Vingt mille lieues sous les mers* (1869). *Twenty thousand leagues under the sea*. Blackie and Son, Ltd. (1959).
27. Wells, H. G. *The world of William Clissold*. Collins (1926).
28. Wells, H. G. *War of the Worlds* Heineman (1898).
29. Jantsch, E. *Technological forecasting in perspective*. O.E.C.D., Paris (1966).
30. Clausewitz, K. Von *On war*. Penguin Books (1968).
31. Kahn, H. *On thermonuclear war*. Princeton University Press (1960).
32. Allen, J. A. *Studies in innovation in the steel and chemical industries*. Manchester University Press (1967).
33. Ansoff, H. I. *Corporate strategy*. Penguin Books (1968).
34. Kahn, H., and Weiner, A. J. *The year 2000*. Collier-Macmillan Ltd. (1967).
35. Mesarovic M., and Pestel, E. *Mankind at the turning point: The Second Report of the Club of Rome*. Hutchinson, London (1975).
36. Trager, J. *The great grain robbery*, Ballantine, New York (1975).
37. Green, J. R. *A short history of the English people*, Vol. 2, p. 873. Macmillan (1924).
38 Gabor, D. *Innovations: Scientific, technological, and social*. Oxford University Press (1970).

CHAPTER 3

1. Davies, D., and McCarthy, C. *Introduction to technological economics*, pp. 8–9. John Wiley and Sons, London (1968).
2. Hwa, J. C-H., *Impressions of China's reconstruction. Chemy Technol.* January, 1975, pp. 7–11.
3. C A–200, *Chem. Age*, 25 July, 1975, S3–S28.
4. Davies, D. S., and Stammers, J. R. The effect of World War II on industrial science. *Proc. R. Soc.* A 342, 505–18 (1975).
5. Chapman, P. *Fuel's paradise: Energy options for Britain*. Penguin Books (1975).

6. Owen, John *Epigrams*.
7. Oliver, J. W. *History of American Technology*. Ronald Press, New York (1956).
8. Robson, W. A. *Nationalised industry and public ownership*, Allen and Unwin Ltd, (1960).
9. Ferneyhough, F. *The history of railways in Britain*. Osprey Publishing Ltd. (1975).
10. Bono, E. de. *The uses of lateral thinking*. Jonathan Cape Ltd. (1967).
11. Kuhn, T. S. *The Structure of scientific revolutions*. University of Chicago Press (1970).
12. Davies, D. S., and McCarthy, C. *Introduction to Technological Economics*, pp. 14–19. John Wiley and Sons, London (1967).
13. Langrish, J. *Wealth from knowledge: A study of innovation based on the 1966 and 1967 Queen's Award to Industry*. Macmillan (1972).
14. Derry, T. K., and Williams, T. I. *A short history of technology*, p. 311–42. Oxford University Press (1960).
15. *County of Los Angeles Profile of Air Pollution* Los Angeles Air Pollution Control District, California (1971).
16. Snow, C. P. *Two cultures*. Cambridge University Press (1964).
17. McCarthy, M. C. *The employment of highly specialised graduates: A comparative study in the U.K. and the U.S.A.* (Science Policy Studies No. 3). H.M.S.O. (1968).
18. *Origins and development of operational research in the Royal Air Force*. Air Force Publication O. 3368 H.M.S.O. (1963).
19. Tomlinson, R. C. *Operational research comes of age: a review of the work of the Operational Research Branch of the National Coal Board, 1948–69*, Tavistock Publications (1971).
20. Rothschild, Lord. *A framework for Government research and development (The organisation and management of government research and development)*. Cmnd. 4814 H.M.S.O., (1971).
21. Swann, M. *The flow into employment of scientists, engineers and technologists (A Report of the Working Group on Manpower for Scientific Growth)*. Cmnd. 3760, H.M.S.O. (1968).

CHAPTER 4

1. Heitler, W. *Elementary wave mechanics with applications to quantum chemistry*. Clarendon Press (1956).
2. Hinshelwood, C. N. *The structure of physical chemistry*. Clarendon Press (1951).
3. Lewis, G. N. The atom and the molecule. *J. Am. chem. Soc.* **38** 762–85 (1916).
4. van't Hoff, J. H. *La chimie dans l'espace*. Roterdam (1874) and *Bull. Soc. chim.* **24**, 295, 338 (1875); le Bec, J. A. Sur les relations qui existent entre les formules atomiques, *ibid.* **22**, 337.
5. Bray, J. *Decision in Government*. Gollancz (1970).
6. Low, D. *The Fearful Fifties*. Bodley Head (1960).

References

7. Beer, S., and Revans, R. W. *Operational research and personnel management.* Institute of Personnel (1959).
8. Jenkins, G. M., and Youle, P. V. *Systems engineering: A unifying approach in industry and society,* pp. 173–98. C. A. Watts and Co. London (1971).
9. Clough, D. J. Lewis, C. G. and Oliver, A. L. (eds.) *Manpower planning models.* English Universities Press (1974).
10. Ferrell, W. R., and Milligan, R. H. *An experimental test of K.S.I.M.* Paper Tu PN1/5. Systems and Industrial Engineering, Department, University of Arizona, Tucson.
11. Davies D. S., and McCarthy, C. *Introduction to technological economics,* pp. 20–41. John Wiley and Sons, London (1968).
12. Roeber, J. *Social change at work: The I.C.I. Weekly Staff Agreement.* Duckworth (1975).
13. Duncanson, L. A., and Youle, P. V. On-line control of olefin plants. *Chem. Process.* 51 (5) 49–52 (1970).
14. Davies, D. S., and McCarthy, C. *Introduction to technological economics,* pp. 71–3. John Wiley and Sons, London (1968).
15. Beenhakker, H. L. Sensitivity analysis of the present vaue of a project, *Eng. Economist,* 20 (2), 123–49 (1975).
16. *Perspectives on experience,* Boston Consulting Group Inc. (1970).
17. Rapoport, A. *Fights, games and debates,* pp. 213–25. University of Michigan Press (1960).
18. *Ibid.,* pp. 107–29.
19. Robbins, Lord. *Higher Education: Report of the Committee appointed by the Prime Minister 1961–63.* Cmnd. 2154. H.M.S.O. (1963).
20. Cole, H. S. D., *et al. Thinking about the future (A critique of* The Limits to growth). Sussex University Press (1973).

CHAPTER 5

1. Cook, E. The flow of energy in an industrial society. *Sci. Am.* 225, 135–44. (1971).
2. Davies, D. S. Energy conservation in the chemical industry. Part II: Economics and politics of the return of coal and cellulose. *Chem. Ind.* 20 Sept. 1975, pp. 771–5.
3. Arnon, I. In *Science policy and development* (Ed. Tal and Ezrachi), p. 119. Gordon and Breach (1972).
4. Hendricks, S. B. *Resources and man,* pp. 65–86. W. H. Freeman and Co. (1969).
5. Weiss, D. Resource full Australia. *Chem. Britain,* 12, 8–15 (1976).
6. Anstey, R. *The Atlantic Slave Trade and British Abolition (1760–1810).* Macmillan (1975).
7. Maddox, J. Raw materials and the price mechanism. *Nature, Lond.* 236, 331–4 (1972).
8. Lovering, T. S. *Resources and Man,* pp. 109–34. W. H. Freeman and Co. (1969).

9. P. Chapman, *Fuels Paradise: Energy options for Britain*, pp. 97–108. Penguin Books (1975).
10. Postgate, J. R. New advances and future potential in biological nitrogen fixation. *J. appl. Bact.* 37 (2) 185–202 (1974).
11. *Annual Abstraction of Statistics 1972*. Central Statistical Office. H.M.S.O. (1972).
12. Cole, H. S. *et al.* Thinking about the future (A critique of The limits to growth), pp. 35–42. Sussex University Press (1973).
13. Faltermayer, E. Metlas: The warning signals are up. *Fortune*, 86 (4), 10 (1972).
14. Fisher, H. A. L. *A history of Europe.* Edward Arnold and Son Ltd. (1936).

CHAPTER 6

1. Mishan, E. J. *The costs of economic growth.* Staples Press (1967).
2. Masters, D. *Miracle drug: The inner history of penicillin.* Eyre and Spottiswoode Ltd. (1946).
3. Schumacher, E. F. *Small is beautiful: A study of economics as if people mattered.* Blond and Briggs Ltd. (1973).
4. Spinks, A. Our changing research. *I.C.I. Magazine*, 52, 122–8 (1974).
5. Zuckerman, Sir Solly. *Beyond the ivory tower: functions of public and private science*, pp. 162 Weidenfeld and Nicholson (1970).
6. Jantsch, E. *Technological forecasting in perspective*, pp. 99–108 O.E.C.D., Paris (1966).
7. Anon. A survey of Britain's savings institutions. *Economist*, 257, 29 Nov. 1975 (supplement, 28 pp).
8. Sandilands, F. E. P. *Inflation accounting (Report of the Inflation Accounting Committee).* Cmnd. 6225 H.M.S.O. (1975).
9. Davies, D. S. Discontinuities in chemistry and chemical technology. The Fifth Royal Society Technology Lecture. *Proc. R. Soc.* A 330, 149–72 (1972).
10. Bray, J. *Decisions in Government*, p. 115 Gollancz (1970).
11. Shaw, G. B. *Man and Superman.* Constable (1903).
12. Fisher, J. C., and Pry, R. H. In *Industrial applications of Technology Forecasting* (Ed. M J. Cetron). John Wiley and Sons, New York (1971).

CHAPTER 7

1. Hinshelwood, C. N. *Chemical kinetics of the bacterial cell.* Clarendon Press, Oxford (1946).
2. Gabor, D. *Innovations: Scientific, technological, and social*, p. 7. Oxford University Press (1970).
3. Mishan, E. J. *The costs of economic growth*, pp. 74–100 Staples Press (1967).
4. Robbins, Lord. *Higher education: report of the Committee appointed by the Prime Minister 1961–63.* Cmnd. 2154. H.M.S.O. (1963).

References

5. Young, M. *Family and kinship in East London*. Penguin Books (1969).
6. Buchanan, C. *Traffic in towns*. Penguin Books (1964).
7. Schein, E. H. *Process consultation: Its role in organisation development*, pp. 22–6. Addison-Wesley, Massachusetts (1969).
8. Tannenhaum, R., and Schmidt, W. H. How to choose a leadership pattern. *Harvard Business Review*, 36, 95–101 (1958).
9. Galbraith, J. K. *The new industrial state. (2nd ed.) Andre Deutsch Ltd. (1972).*
10. *Perspectives on experience*, pp. 45–7. Boston Consulting Group Inc. (1970).
11. Rapoport, A. *Fight, games and debates*, pp. 107–29. University of Michigan Press (1960).

CHAPTER 8

1. Hartmann, H. F., and Norman, N. *Nature in the balance*. Heinemann Educational Books (1973).
2. Warner, Sir F. *Management in the scienced based industries (Design and location of plant)*, pp. 78–9. Symposium held in Dublin, April 1968. Royal Insititue of Chemisty (1968).
3. Hardie, D. W. F. *A History of the chemical industry in Widnes*. I.C.I. Ltd. (General Chemicals Division) (1950).
4. Clayton, R. J. Only connect. *Proc. Instn. elec. engrs.* 123 (1), 1–12 (1976).
5. Blanco-White, T. A. *Patents for inventions*, 4th edn. Stevens and Sons (1974).
6. *Industrial Relations Act 1971*, Chapter 72. H.M.S.O. (1972).
7. *Air pollution control: an intregrated approach*. Royal Commission on Environmental Pollution (5th Report). H.M.S.O. (1976).

Index

The technologist today faces a wider variety of problems than ever before and in addition to his traditional skills he needs an understanding of sociology, politics, and psychology. The authors here put forward from their extensive experience a comprehensive approach in terms of modelling for the exploration of alternative policies. This powerful method of investigation is considered in relation to present and future concerns in natural resources and energy, capital and finance, human behaviour and organization, the environment, and the law. Thus armed, the complete technologist will be able to make a valuable contribution to society.

The book is written very much in terms of action as well as understanding. It should be of interest to a wide variety of readers, non-scientists as well as scientists, including senior industrialists, administrators, and civil servants, as well as to the technologists themselves.

The Club of Rome has always recognized the fundamental role which a wisely oriented and humane technology could play in helping to resolve the tangle of interacting problems of contemporary society which we term the 'world problematique'. This is a wise, witty and important book which considers how such a role might be developed within the political, economic and cultural constraints of tomorrow's society. Its basic justification is well expressed in the understatement of the first sentence, 'The study of people does not feature prominently in technological education.'

Alexander King
Aurelio Paccei
Co-founders of the Club of Rome